EIT Mechanical Review

Revised Edition

Lloyd M. Polentz, P.E.

Engineering Press **Austin, Texas**

ISBN 1-57645-024-4

Engineering Press
P.O. Box 200129
Austin, TX 78720-0129

(800)800-1651
FAX (800)700-1651

Contents

Preface

The Fundamentals of Engineering/Engineer-In-Training (FE/EIT) exam has dramatically changed. For about the last fifty years the 8-hour exam has been a test of a dozen topics that are the core of every engineer's education: Math-Chem-Physics-Statics-Dynamics-Fluids-Thermo-Circuits-Engr Econ-Mat Sci-Computers and Mech of Materials. All that changed in the Fall of 1996. Now the exam has two distinct parts to it. The morning four-hour portion is pretty much like the old exam with about the same dozen topics. This book has been revised in this edition to closely follow the afternoon topics and the afternoon exam content of the *Fundamentals of Engineering Reference Handbook.*

The afternoon four-hour portion of the exam is new and different. In fact there are six different afternoon exams. Each one covers a different branch of engineering: civil, mechanical, electrical, chemical, industrial and general. While the morning exam covers the first 2-2½ years of an engineer's education, the afternoon exam covers the last 1½-2 years of an engineers education in his specific branch of engineering. This means that an orderly review in preparation for the exam has two components and requires two books. Everyone can use the *Engineer-In-Training License Review* book for the morning exam, as can general engineers for the afternoon exam. All other engineers will need one of the following:

EIT Civil Review
EIT Mechanical Review
EIT Electrical Review
EIT Chemical Review
EIT Industrial Review

If you encounter any errors in this book, we would appreciate being notified. Notes and comments may be sent directly to me, c/o Engineering Press, P.O. Box 200129, Austin, TX 78720-0129.

These books, in their previous editions, have helped tens of thousands of engineers prepare for and pass the FE/EIT exam. The team of engineering professors and I have worked diligently to provide you the best possible review to quickly and efficiently prepare for this changed exam. We wish you good luck in the exam.

Lloyd M. Polentz

Selected books of interest published by Engineering Press

LICENSE REVIEW BOOKS

For the FE/EIT Examination--2 Books are needed:

Engineer-In-Training License Review

and One of these supplements:

EIT Civil Review
EIT Mechanical Review
EIT Electrical Review
EIT Chemical Review
EIT Industrial Review

For the PE Examinations:

Civil Engineering License Review
Civil Engineering License Problems and Solutions
Environmental Engineering Problems and Solutions
Seismic Design of Buildings and Bridges
Mechanical Engineering License Review
Electrical Engineering License Review
Electrical Engineering License Problems and Solutions
A Programmed Review for Electrical Engineering
Chemical Engineering License Review
Structural Engineer Registration: Problems and Solutions

EXAM FILES

Calculus I
Calculus II
Calculus III
College Algebra
Circuit Analysis
Differential Equations
Dynamics
Engineering Economic Analysis
Fluid Mechanics
Linear Algebra
Materials Science
Mechanics of Materials
Organic Chemistry
Physics I—Mechanics
Physics III—Electricity and Magnetism
Probability and Statistics
Statics
Thermodynamics

OTHER BOOKS

Being Successful As An Engineer
The Basics of Structural Analysis
Fluid Mechanics
Engineering Economic Analysis
Elements of Bioenvironmental Engineering
Soil Mechanics Laboratory Manual

Free Catalog

For a free catalog describing the books above and 80 other license review books, with prices and ordering information,

write to:

Engineering Press Bookstore
P.O. Box 200129
Austin, TX 78720-0129

Phone: (800) 800-1651
FAX: (800) 700-1651

INTRODUCTION

BECOMING A PROFESSIONAL ENGINEER

To achieve registration as a Professional Engineer, there are four distinct steps: 1) education, 2) the Fundamentals of Engineering/Engineer-In-Training (FE/EIT) exam, 3) professional experience, and 4) the professional engineer exam. These steps are described in the following sections.

Education

Generally, no college degree is required to be eligible to take the FE/EIT exam. The exact rules vary, but all states allow engineering students to take the FE/EIT exam before they graduate, usually in their senior year. Some states, in fact, have no education requirement at all. One merely need apply and pay the application fee. Perhaps the best time to take the exam is immediately following completion of related coursework. For most engineering students, this will be in the senior year.

Fundamentals of Engineering/ Engineer-In-Training Examination

This eight-hour, multiple-choice examination is known by a variety of names: Fundamentals of Engineering, Engineer-In-Training (EIT.), or Intern Engineer, but no matter what it is called, the exam is the same in all states. It is prepared and graded by the National Council of Examiners for Engineering and Surveying (NCEES).

Experience

States that allow engineering seniors to take the FE/EIT exam have no experience requirement. These same states, however, generally will allow other applicants to substitute acceptable experience for coursework. Still other states may allow one to take the FE/EIT exam without any education or experience requirements.

Professional Engineer Examination

The second national exam is called Principles and Practice of Engineering by NCEES, but many refer to it as the Professional Engineer exam or P.E. exam. All states, plus Guam, the District of Columbia, and Puerto Rico use the same NCEES exam. Typically, four years of acceptable experience is required before one can take the Professional Engineer exam, but the requirement may vary from state to state. Review materials for this exam are found in other engineering license review books.

FUNDAMENTALS OF ENGINEERING/ ENGINEER-IN-TRAINING EXAMINATION

Background

Laws have been passed that regulate the practice of engineering to protect the public from incompetent practitioners. Beginning about 1907 the individual states began passing *title* acts regulating who could call themselves an engineer. As the laws were strengthened, the *practice* of engineering was limited to those who were registered engineers, or to those working under the supervision of a registered engineer. Originally the laws were limited to civil engineering, but over time they have evolved so that the *titles* and sometimes the *practice* of most branches of engineering are included. There is no national registration law; registration is based on individual state laws and is administered by boards of registration in each state.

Examination Development

Initially, the states wrote their own examinations, but beginning in 1966 the NCEES took over the task for some of the states. Presently the NCEES exams are used by all states. Now it is easy for engineers that move from one state to another to achieve registration in the new state. About 50,000 engineers take the FE/EIT exam annually. This equals about half the engineers graduated in the U.S. each year.

The development of the FE/EIT exam is the responsibility of the NCEES Committee on Examinations for Professional Engineers. The committee is composed of people from industry, consulting, and education, some of whom are subject-matter experts. The test is intended to evaluate an individual's understanding of basic sciences as well as engineering sciences along with one's ability to apply this knowledge to engineering problems. Every five years or so, NCEES conducts an engineering task analysis survey. People in industry, consulting, and education are surveyed to review engineering practice and the knowledge needed for it. From this, NCEES develops what they call a "matrix of knowledge" that forms the basis for the FE/EIT exam structure described in the next section.

The actual exam questions are prepared by the NCEES committee members, subject matter experts, and other volunteers. All people participating must hold professional registration. When the questions have been written, they are circulated for review in workshop meetings and by mail. Both single- and multi-part questions are used; the latter are arranged so the solution of each succeeding part does not depend on the correct solution of a previous part. All problems are four-way multiple choice.

Examination Structure

The FE/EIT exam is divided into a morning four-hour section and an afternoon four-hour section. The Fundamentals of Engineering Examination in the afternoon period has changed to a discipline specific examination. Six exams are prepared in the afternoon test booklet, one for each of the following five disciplines: Civil, Mechanical, Electrical, Chemical, Industrial; and a General topic for those examinees not covered by the five engineering disciplines. The test consists of 60 problems in the topic. This may be of great benefit to those specializing in one of the five major engineering disciplines. Most of the test's topics cover the third and fourth year of your college courses. These are the courses which you will use for the balance of your engineering career, so the test becomes focused to your own needs. Graduate engineers will find the PM discipline test to their advantage as the broad fundamentals test usually causes many engineers to do a good deal of review of the earliest class-work.

The major subjects for the afternoon sections are:

Civil	No.Probs.
Soil Mechanics & foundations	6
Structural Analysis, frames, trusses	6
Hydraulics & hydro systems	6
Structural Design, concrete, steel	6
Environmental Engineering, waste water, solid waste treatment	6
Transportation facilities, highways railways, airports	6
Water purification and treatment	6
Surveying	6
Computer and Numerical methods	6
Legal & professional aspects	3
Construction Management	3

Mechanical	
Mechanical Design	6
Dynamic Systems,vibration,kinematics	6
Thermodynamics	6
Heat Transfer	6
Fluid Mechanics	6
Stress Analysis	6
Measurement and Instrumentation	6
Material Behavior/ Processing	3
Computer, Automation, Robotics and Numerical methods	3
Energy Conversion and Power Plants	3
Automatic control	3
Refrigeration and HVAC	3
Fans, Pumps, and Compressors	3

Electrical	
Digital Systems	6
Analog Electronic Circuits	6
Electromagnetic Theory & Applications	6
Network Analysis	6
Control Systems Theory and Analysis	6
Solid State Electronics and Devices	6
Communications Theory	6
Signal Processing	3
Power Systems	3
Computer Hardware Engineering	3
Computer Software Engineering	3
Instrumentation	3
Computer and Numerical Methods	3

Chemical	No.Probs
Material & Energy Balances	9
Chemical Thermodynamics	6
Mass Transfer	6
Chemical Reaction Engineering	6
Process Design and Econ Evaluation	6
Heat Transfer	6
Transport Phenomenon	6
Process Control	3
Process Equipment Design	3
Computer and Numerical Methods	3
Process Safety	3
Pollution Prevention	3

Industrial	
Production Planning & Scheduling	3
Engineering Economics	3
Engineering Statistics	3
Statistical Quality Control	3
Manufacturing Processes	3
Mathematical Optimization-Modeling	3
Simulation	3
Facility design and location	3
Work Performance and Methods	3
Manufacturing Systems Design	3
Industrial Ergonomics	3
Industrial Cost Analysis	3
Material Handling System Design	3
Total Quality Management	3
Computer Computations-Modeling	3
Queuing Theory and Modeling	3
Design of Industrial Experiments	3
Industrial Management	3
Information System Design	3
Productivity Measurement	3

General	
Mathematics	12
Electrical Circuits	6
Statics	6
Thermodynamics	6
Chemistry	4
Dynamics	4
Mechanics of Materials	4
Material Science	3
Computers	3
Ethics	3
Engineering Econ	3

Taking the Examination

Examination Dates

The National Council of Examiners for Engineering and Surveying (NCEES) prepares FE/EIT exams for use on a Saturday in April and October each year. Some state boards administer the exam twice a year; others offer the exam only once a year. The scheduled exam dates are:

	April	October	November
1998	25	31	—
1999	24	30	—
2000	15	28	—

Those wishing to take the exam must apply to their state board several months before the exam date. We recommend that Civil, Mechanical, Electrical, Chemical, and Industrial engineers take the discipline specific exam. All others should take the General examination.

Examination Procedure

Before the morning four-hour session begins, the proctors will pass out exam booklets and a scoring sheet to each examinee. An HB or #2 pencil is to be used to record answers. Space is provided on each page of the examination booklet for scratchwork. The scratchwork will *not* be considered in the scoring.

The examination is closed book. You may not bring any reference materials with you to the exam. To replace your own materials, NCEES has prepared a *Fundamentals of Engineering (FE) Reference Handbook.* The handbook contains engineering, scientific and mathematical formulas and tables for use in the examination. Examinees will receive the handbook from their state registration board prior to the examination. The *FE Reference Handbook* also is included in the exam materials distributed at the beginning of each four-hour exam period. If you do not already have a copy of the *FE Reference Handbook* you should purchase one right away.

There are three versions (A, B, and C) of the exam. These have the major subjects presented in a different order to reduce the possibility of examinees copying from one another. The first subject on your exam, for example, might be fluid mechanics, while the exam of the person next to you may have electrical circuits as the first subject.

The afternoon session begins following a one-hour lunch break. The afternoon exam booklets will be distributed along with a scoring sheet. At the beginning of the afternoon test period, the examinee will mark the answer sheet as to which exam he is taking. There will be (60 multiple choice questions, each of which carries twice the grading weight of the morning exam questions.

If you answer all questions more than 30 minutes early, you may turn in the exam materials and leave. In the last 30 minutes, however, you must remain to the end of the exam period to ensure a quiet environment for all those still working, and to ensure an orderly collection of materials.

Examination-Taking Suggestions

Those familiar with the psychology of examinations have several suggestions for examinees:

1. There are really two skills that examinees can develop and sharpen. One is the skill of illustrating one's knowledge. The other is the skill of familiarization with examination structure and procedure. The first can be enhanced by a systematic review of the subject matter. The second, exam-taking skills, can be improved by practice with sample problems-that is, problems that are presented in the exam format with similar content and level of difficulty.

2. Examinees should answer every problem, even if it is necessary to guess. There is no penalty for guessing. The best approach to guessing is to first eliminate the one or two obviously incorrect answers among the four alternatives. If this can be done, the chance of selecting a correct answer obviously improves from 1 in 4 to 1 in 2 or 3.

3. There are 60 points possible in the afternoon and you will have four hours or 240 minutes. Subtract say 30 minutes to get organized, *read the instructions*, go to the restroom, review the examination, etc. That leaves 210 minutes. When you review the examination mark those problems which you think you can do without too much trouble with a plus sign, those that you think you might not be able to handle with an X, and the others with a check. You have 3½ minutes per point available to you. Do the easy (to you) problems first, then work on those with a check mark, and lastly work at those you've marked with an X.

For a two-point problem you have available seven minutes, don't spend more time than that on any two-point problem, or, proportionately on any other problem. There is always a feeling of challenge, the "I know I can do it if I just take a bit more time" feeling. Don't do it, the chances are very good that you will spend enough time trying to satisfy your ego that you will be invited back to take the exam the next time it is given. Your goal is to get as many points as you can in the allotted time and easy points are just as valuable as difficult (to you) points. So keep half an eye on your watch, don't make a big deal out of it, but if you find yourself running over the length of time you think should be allotted, drop it and go on to another problem you think you can do. If you have spare time at the end of the exam period, then go back and take another whack at the problem you are "sure that you can do".

4. Read all four multiple choice answers before making a selection. The first answer in a multiple choice question is sometimes a plausible decoy-not the best answer.

5. Do not change an answer unless you are absolutely certain you have made a mistake. Your first reaction is likely to be correct.

6. If time permits, check your work.

7. Do not sit next to a friend, a window, or other potential distraction.

PREPARATION FOR THE EXAMINATION

One recommendation made consistently by those who have taken the EIT examination is "work a lot of problems". Many engineers a few years out of school have grown very rusty in setting up problems, and then working them. If you haven't refreshed your memory on how to set up a problem to work it logically before you go to the examination room--it may be to late.

License Review Books

There are two rules that we suggest you follow in selecting license review books to ensure that you obtain up-to-date materials:

1. Consider only the purchase of materials that have a 1993 or later copyright. Prior to October, 1993 reference books could be brought to the exam, and some review books added unnecessary reference materials to get them accepted as reference books. That strategy was successful then, but it makes no sense now. Nothing can be taken into the exam, and the unnecessary material in these older books diverts attention from the exam content. Look for books that cover what you need to know--no more, and no less.

2. Even if a license review book has a recent copyright date, is the content up to date? Books that ignore the multiple-choice format, or include irrelevant material, are not up to date.

Textbooks

If you still have your university textbooks, they can be useful in prepairing for the exam, unless they are too out of date. To a great extent the books will be like old friends with familiar notation. You probably need both textbooks and license review books for efficient study and review. You must, however, pay close attention to the *FE Reference Handbook and* the notation used in it, because it is the only book you will have in the exam.

Examination Day Preparations

There is no doubt that the exam day will be a stressful and tiring one. You should take steps to eliminate the possibility of unpleasant surprises. If at all possible, visit the examination site ahead of time. Try to determine such items as:

1. How much time should I allow for travel to the exam on that day. Plan to arrive about 15 minutes early. That way you will have ample time, but not too much time. Arriving too early and mingling with others who also are anxious, can increase your anxiety and nervousness.

2. Where will I park?

3. How does the exam site look. Will I have ample workspace? Will it be overly bright (sunglasses) or cold (sweater), or noisy (earplugs)? Would a cushion make the chair more comfortable?

4. Where is the drinking fountain? Lavatory facilities? Pay phone?

5. What about food? Should I take something, along for energy in the exam? A light bag lunch during the break probably makes sense.

Items to Take to the Examination

● Calculator-NCEES says you may bring a battery-operated, silent, non-printing calculator. You need to determine whether or not your state permits pre-programmed calculators. Bring extra batteries for your calculator just in case; and many people feel that bringing a second calculator is also a very good idea.

● Clock-You must have a time plan and a clock or wristwatch.

● Pencils-You should consider using mechanical pencils so you don't have to run around sharpening pencils. And you surely do not want to drag along a pencil sharpener.

● Eraser-Try a couple to decide what to bring along. You must be able to change answers on the multiple choice answer sheet, and that means a good eraser.

● Exam Assignment Paperwork-Take along the letter assigning you to the exam at the specified location to prove that you are the registered person. Also bring something with your name and picture (driver's license or identification card).

● Items Suggested By Your Advance Visit--If you visit the exam site, it probably will suggest an item or two that you need to add to your list.

● Clothes-Plan to wear comfortable clothes. You probably will do better if you are slightly cool, so it is wise to wear layered clothing.

Examination Scoring

The questions are machine-scored by scanning. The answer sheets are checked for errors by computer. Marking two answers to a question, for example, will be detected and no credit given.

Passing the FE/EIT Examination

The 120 morning and 60 double-weighted afternoon problems yield a raw score of 240. The passing score is set as the equivalent of a raw score of 124 out of 280 on the October, 1990 examination. To translate this established level of difficulty to new examinations, a portion of each exam contains problems given on previous exams. An analysis of this subset of problems can be used to scale the new exam raw score to the same level of difficulty as 124 on the October, 1990 exam. This produces a standardized exam where the percentage of applicants passing, may vary from exam to exam, but they all demonstrate an established level of competency. Since most state licensing laws require 70 on the exam to pass, the raw passing score (equivalent to 124 on the October, 1990 exam) is converted to 70 and reported to the states accordingly. Higher raw scores are reported proportionally higher than 70; lower raw scores as proportionally lower than 70. What does it all mean? You are faced with 120 single-weighted morning problems and 60 double-weighted afternoon problems which makes a maximum possible score of 240. You must correctly answer problems with an equivalent score of about 124 to receive a passing score. So, you do *not* need to answer 70% of the problems correctly; you will get a converted score of 70 if you answer about 45% of the problems correctly.

The converted scores are reported to the individual state boards in about two months, along with the recommended pass or fail status of each applicant. The state board is the final authority of whether an applicant has passed or failed the exam.

Although there is some variation from exam to exam, the following gives the approximate passing rates:

Applicant's Degree	Percent Passing Exam
Engineering from accredited school	62%
Engineering from non-accredited school	*50*
Engineering Technology from accredited school	42
Engineering Technology from non-accredited school	33
Non-Graduates	36
First-time Applicants	67
All Applicants	56

If you do not pass the exam on your first attempt, you can re-apply and take it again.

THIS BOOK

This book has been organized to cover the afternoon portion of the Fundamentals of Engineering/Engineer-In-Training (FE/EIT) examination, with nothing added and hopefully nothing left out. Each of the topic authors has created a chapter that reflects the essential material at the level of the FE/ EIT exam.

Chapter 1
Mechanical Design

Mechanical Design, or Machine Design as it is often called, combines all of the disciplines in the study of Mechanical Engineering. A typical design problem requires first the development of a mechanism to perform the desired function. Next it has to be demonstrated that the device will perform without failing and without imperilling the user or the public. Attention should also be paid to the cost of the device including material, manufacturing, operating, and maintenance costs.

EXAMPLE 1-1

Design a square key for a gear wheel attached to a 2.5 cm diameter shaft to transmit 10.0 kw of power at a speed of 1750 rpm. The gear wheel is 1.5 cm thick. If the key is to be made of steel with an allowable shear stress of 200 MPa, what size key should be used if a safety factor of three is required assuming shear stress controls?

Solution:

The allowable stress on the key material would equal
200/3 = 66.67 MPa.
10.0 kW = 10.0 kJ/s and a joule is a N·m.
The shaft rotates at a speed of 1,750/60 = 29.17 rev/s
and transmits energy at the rate of: (force × m/s)
$F \times \pi \times 0.025 \times 29.17 = 2.2910 \times F$ N·m/s or
$2.2910 \times F$ watts where F = force on key in newtons.
$10{,}000$ watts/$(2.2910 \times F)$ $F = 4{,}364.9$ N
required shear area of key = $4{,}365/66.67 \times 10^6$
$A = 65.47 \times 10^{-10}$ m²
The required key width would = A/0.015 = 4.36 mm■

Interference Fit, some times it is desirable to provide a force fit where the O.D. of the shaft is larger than the I.D. of the mating part. For small interferences this may be accomplished by forcing the shaft into the smaller diameter hole. For larger interferences it will be
necessary to expand the hole diameter and/or reduce the shaft diameter by heating or cooling as, for example, shrinking rims onto railroad car wheels.

EXAMPLE 1-2

If a car wheel is 60 cm in diameter and a rim is 0.9 mm smaller in I.D. to what temperature would it be necessary to heat the rim to provide a 0.100 mm clearance to assemble the rim on the wheel? Both members are made of steel. Dry ice at -79 C is available to cool the wheel. The coefficient of thermal expansion for steel is 11.7×10^{-6} m/(m·C).

Solution:

The total ΔD required is 1.00 mm. For an ambient temperature of 20 C, the wheel diameter would reduce:

$600 \times 99 \times 11.7 \times 10^{-6} = 0.6950$ mm
The rim diameter would then have to expand 0.3050 mm
$\Delta T = 0.3050/(600 \times 11.7 \times 10^{-6}) = 43.4$ C
The rim would have to be heated to 63.4 C ■

To examine this a little further, assume all deformation took place in the rim after assembly i.e. the wheel was strong enough not to deform after the rim was installed. What would be the stress in the rim after it had been shrunk onto the wheel?

$\Delta D = 0.90$ mm Circumference = πD so $\Delta C = \pi x 0.90$ mm

$\delta_C = 0.90\pi/(600\ \pi) = 0.0015$ m

stress = $E \cdot \delta = 2.1 \times 10^{11} \times 0.0015 = 315$ MPa ■

EXAMPLE 1-3

A 10 kw motor operates at a speed of 1,750 rpm. What diameter shaft should be used to transmit this power using a steel with a tensile strength of 572 MPa and a tensile yield strength of 496 MPa? Use a safety factor of 2.5, and assume the shear yield strength of the shaft steel is 0.60 × yield strength in tension.

Solution:

Shear stress = $Tq \cdot c/J$

where $J = \pi D^4/32$, the polar moment of inertia and $c = D/2$ so $S_s = 16 \cdot Tq/\pi \cdot D^3$

Watt = joule/sec = N·m/s 1,750 rpm = 29.17 rev/s

Allowable shear stress = 496 × 0.60/2.5 = 119.04 MPa

Tq = 10,000 watts/(29.167 rev/sec × 2·π) = 54.57 joules

$D = [16 \times Tq/(S_s \cdot \pi)]^{1/3} = 0.0133$ m or 1.33 cm ■

If it is desired to use a 2.00 cm O.D. hollow shaft, what would be the size of the bore to produce the same shear stress in the hollow shaft?

Since the stress is directly proportional to the polar moment of inertia, the J of the hollow shaft would have to be the same as for the solid shaft calculated above.

$J = \pi \cdot D^4/32$ since $J_1 = J_2$ $D^4 = D_{OD}^4 - D_{ID}^4$

$1.33^4 = 2^4 - D_{ID}^4$ $D_{ID} = (16 - 3.129)^{1/4} = 1.894$ which gives a wall thickness of 0.529 mm. ■

EXAMPLE 1-4

In the above example a 1.500 cm solid shaft was selected. The power from the motor was delivered to the driven machine through a 30.00 cm diameter pulley. If the pulley were attached to the shaft at a distance of 15.00 cm from a bearing support, what would be the bending stress in the shaft (a), and what would be the total shear stress, τ, (b)?

Solution:

The torque output of the motor is 54.57 J (N·m). The force acting on a belt at a distance of 30.00/2 cm would equal (1.00/0.150) × 54.57 = 363.8 N. The maximum bending moment would equal 0.15 × 363.8 = 54.57 J

$S = M \cdot c/I$ where I is the transverse moment of inertia, and since $J = I_x + I_y$ it will equal one-half of the polar moment of inertia for a circular section.

$I_x = \pi D^4/64 = 2.485 \times 10^{-9}$ m^4

$S = (54.57 \times 0.015/2)/(2.485 \times 10^{-9})$

S equals 164.7 MPa, bending stress (a) ■

$J = 2 \times I_x = 4.970 \times 10^{-9}$

$S_s = (54.57 \times 0.0150/2)/(4.970 \times 10^{-9}) = 82.35$ MPa

Maximum shear stress $\tau = [s_s^2 + (S/2)^2]^{1/2} = 117.1$ MPa ■

N.B. The relationship for the maximum shear stress can be derived from the diagram of Mohr's Circle given in the FE Handbook, using only two tensile stresses, i.e. let $\sigma_2 = 0$. It would be a good idea for an examinee to study the section on Mohr's Circle before the exam if not already acquainted with it.

COMPOSITE BEAM

How much stiffer would an aluminum beam 5.00 cm wide by 16.00 cm deep be if the bottom 1.00 cm were replaced by steel? See Fig 1-1.

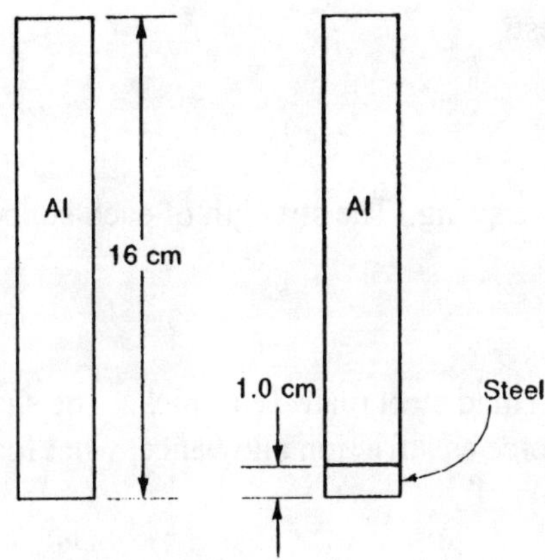

Figure 1-1.

$E_S = 2.1 \times 10^{11}$ Pa $\qquad\qquad E_{Al} = 6.9 \times 10^{10}$ Pa

$E_S/E_{Al} = 3.0435$ so the equivalent all aluminum beam would be a top section 5.00 cm by 15.00 cm with a 1.00 cm by 15.218 cm flange section at the bottom.

The stiffness of a beam is proportional to its moment of inertia. $I = b \cdot h^3/12$ for a rectangular section. For the initial all aluminum beam:

$$I = 5 \times 16^3/12 = 1{,}706.67 \text{ cm}^4$$

If the bottom 1.00 cm were replaced by a steel strip, the location of the neutral axis would shift. Taking area moments about the top of the beam:

$$5 \times 15 \times 7.5 + 1 \times 15.218 \times 15.5 = (75 + 15.218) \times y$$

$y = 798.379/90.218 = 8.849$ cm from the top of the beam is the location of the neutral axis of the composite beam.

The new I, I_c, can be determined by means of the parallel axis theorem.

$$I_c = 5 \times 15^3/12 + (75 \times 1.346^2)$$
$$+ 15.218 \times 1^3 + (15.218 \times 6.654^2) = 2{,}217.19 \text{ CM}_4$$
$$2{,}217.19/1{,}706.67 = 1.30$$

So the composite beam would be 30% stiffer.

EXAMPLE 1-5

A swing is to be designed for a public park. The support beam will be 3.00 m high, and the swing will be designed to hold a 100 kg person. What should be the strength of the lines holding a swing seat? Since this is a public park, it has to be assumed that the swing will be subjected to uses (or mis-uses) far beyond what any one might imagine, so a safety factor of five has been decided upon.

Solution:

The maximum load on the swing seat will occur at the bottom of the arc and will equal the gravitational force exerted by the swinger and the centrifugal force, F_c.

$F_c = m \cdot v^2/R$

Assume the swinger goes as high as the support beam, ΔP.E. = ΔK.E.

then $v = \sqrt{(2gh)}$ or

$v^2 = 2 \times 9.8066 \times 3.0 = 58.84\ m^2/s^2$

assuming the swing seat is at the level of the ground when at rest.

$F_c = 100 \times 58.84/3 = 1{,}961$ N

The total force down would equal

$1{,}961 + 100 \times 9.8066 = 2{,}942$ N

Assume that a swinger would hold onto only one side of the swing. The strength of each support line should equal $5 \times 2{,}942 = 14{,}710$ N. ■

EXAMPLE 1-6

A punch press is to be designed to punch 2.50 cm diameter holes in mild steel plate 1cm thick. The shear strength of the steel is 250 MPa. If 15% is added to the shearing force as a friction allowance, what force must the punch exert?

Solution:

Shear area = $0.025 \times \pi \times 0.01 = 0.0007854\ m^2$

Force = $250 \times 10^6 \times 785.4 \times 10^{-6} = 196.4$ kN

Add allowance for friction

$F = 1.15 \times 196.4 = 225.9$ kN ■

How much energy must be expended to punch one hole?

Energy = F × distance = 225.9 kN × 0.01 m = 2,259 J ■

The design calls for the punch press to be operated by a 400 watt motor running at a speed of 1,750 rpm. The motor will drive a fly wheel through a hydraulic clutch which is 85% efficient. Energy from the fly wheel will be used to provide the required power to operate the press. The requirement is that the fly wheel should not slow down by more than 20% to punch a hole. What should the moment of inertia of the fly wheel be if rotates at a speed of 250 rpm before its energy is used to operate the punch press?

K.E. of a rotating fly wheel equals $\frac{1}{2} I\omega^2$

$\omega_1 = (250/60) \times 2\pi = 26.180$ rad/s

$\omega_2 = (200/60) \times 2\pi = 20.944$ rad/s

$\Delta K.E. = \frac{1}{2} \cdot I \cdot (685.4 - 438.7) = 246.7 \times I = 2{,}259$ J

$I = 9.157\ kg \cdot m^2$ ($N \cdot m \cdot sec^2$) ■

If a 65.0 cm diam, steel fly wheel of constant thickness is used, how thick should it be? The sp. gr. of steel is 7.85.

$I = \frac{1}{2} MR^2 = 9.157$ R = 0.325 m M = 173.39 kg

The density of steel = 7,850 kg/m³

so the volume of the fly wheel should be 0.0221 m³

The area = 0.3318 m² thickness = 6.661 cm ■

MECHANICAL SPRINGS

The commonest type of spring is the helical spring, made with round wire. Such springs are made to withstand either tensile or compressive loads. The spring rate is a function of the number of active coils, wire diameter, spring diameter, and the torsional modulus of the wire material. For steel wire the torsional or shear modulus--

$G_s = 8.3\times10^{10}$ Pa (See Reference Handbook)

The apparent stress in the wire equals $S_t = 8\cdot F\cdot D/(\pi d^3)$ equals torsional stress. D = mean diameter. However this relationship does not include the effect of the curvature of the wire which increases the stress in the wire. The increased stress is determined by a factor (the Wahl factor, K) which equals--

$K = [(4C-1)/(4C-4)+(0.615/C]$

where C = (Mean coil diameter)/(wire diameter)

EXAMPLE 1-7

A coil spring 4.0 cm in diameter made of 5.0 mm wire is to support a mass of 45 kg. What is the stress in the wire?

Solution:

The load would equal $45.0 \times 9.807 = 441.3$ N

$D = 0.040 - 0.005 = 0.035$ m

Apparent $S_t = 8\cdot0.035\cdot441.3/(\pi\cdot0.125\times10^{-6}) = 314.6$ MPa

$C = 0.035/0.005 = 7 \quad K = 27/24 + 0.615/7 = 1.21$

the actual maximum stress $= 314.6 \times 1.21 = 380.7$ MPa ■

DEFLECTION

The deflection, Δ, of a constant diameter circular spring is given by the relationship

$\Delta = 8FD^3No/Gd^4$

where F = deflection and No = number of active coils

EXAMPLE 1-8

For the spring in the previous example: A compression spring 4 cm diam. made of 5 mm diameter wire, determine the number of active coils required for the spring to deflect 6.0 cm under the load of 45 kg.

Solution:

The calculation to determine the deflection of the spring with D =0.035 m and d = 0.005 m

$\Delta = 0.060 = 8\times441.3\times0.035^3\times No/(8.3\times10^{10}\times0.005^4)$

$3.113 = 0.151\times No$ or number of active coils = 20.6 say 21

For a compression spring the ends would probably have squared and ground ends giving one "dead" coil at each end, so the total number of coils would equal 23. ■

SPRING LOAD

The load on a bolt supplied by a spring plus an additionally applied load will equal the load due to the force of the spring plus the additional load until the bolt has stretched (increased in length) enough to relieve the spring load. Another way of stating this is that the force acting on a bolt which is loaded by a spring plus an applied load equals the force applied by the spring before the additional load was applied plus the force of the applied load minus the reduction in the force applied by the spring due to the elongation of the bolt. When the assembly consists of an ordinary coiled spring the strain of the bolt due to the additionally applied load will be small and can usually be neglected. However, when the "spring" load is supplied by a composition gasket or a copper or aluminum washer or gasket, the reduction in the force applied by the spring can be appreciable.

EXAMPLE 1-9

An eye-bolt support is attached to a rigid steel beam as shown in Fig. 1-2. A very stiff compression spring is used to preload the eye-bolt. The spring is loaded by means of a nut screwed down on the end of the eye-bolt until it exerts a force of 5,000 N. What would be the force exerted on the shank of the bolt if a mass of 500 kg were attached to the eye-bolt?

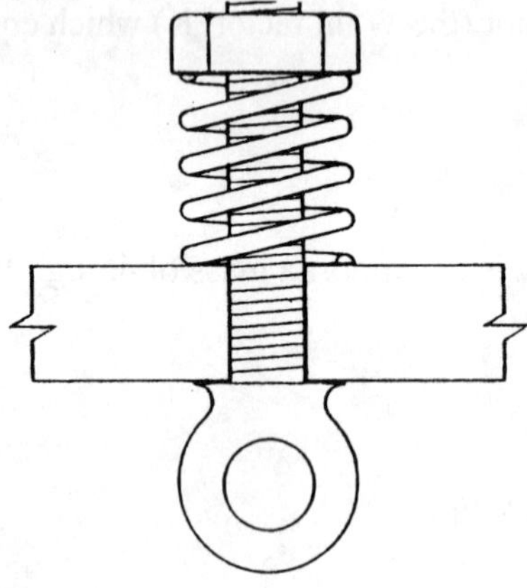

Figure 1-2

Solution:

The force of gravity acting on the support by a mass of 500 kg would equal (500 × g) or $F = 500 \times 9.8066 = 4{,}908.3$ N

This is less than the upward force exerted by the spring, so the load held by the bolt would equal 5.0 kN. ■

What would be the load held by the bolt if a mass of 800 kg were attached to the eye of the bolt?

$F = 800 \times g = 7{,}845$ N ■

This is greater than the initial load on the spring, so the spring would compress farther, and the load on the bolt would be 7.85 kN.

When a bolt is tightened (pre-loaded) against a "soft" substance such as is shown in Fig. 1-3, and a load "P" is applied, it will increase the load on the bolt, but not by the amount of the newly applied load. Stress and strain are proportional, the length of the bolt will increase, but the compression of the "soft" material will decrease by the same amount that the length of the bolt increases. The commonest example of this is the case of a soft seal wherein the gasket material is compressed to seal the joint. The definition of a "soft" material is any material that has a lower coefficient of elasticity than that of the bolt, which is usually steel.

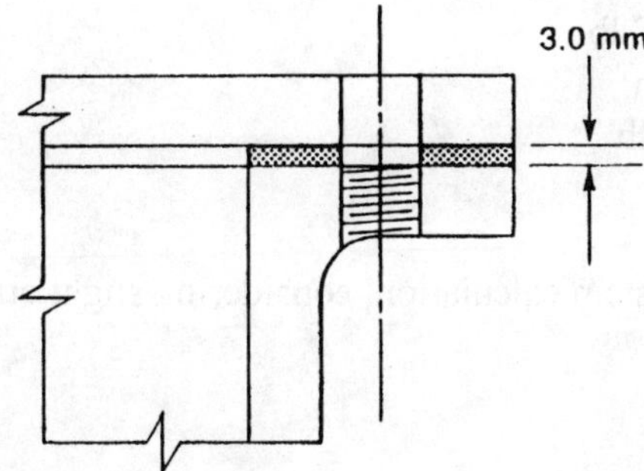

Figure 1-3

EXAMPLE 1-10

Assume that a soft gasket which has a modulus of elasticity equal to 10.0 x 10^8 Pa is used to seal a pressure vessel as shown in the figure. The net area of the gasket (subtracting area of bolt holes) is 88 cm^2 and its thickness is 3.0 mm. The gasket is compressed by 12 steel bolts 15 cm long and 1.0 cm in diameter. Assume the joint has been carefully made up and that each bolt holds exactly 1/12 of the applied load. The preload on each bolt is 11.0 kN. If the force added to the joint due to pressurization of the vessel equals 54.0 kN, how much would the force acting on each bolt increase?

Solution:

The bolt would strain an additional amount to withstand the added load, but the force exerted by the gasket would be reduced since the gasket would expand as the bolt lengthened. Both the bolt and the gasket can be treated as springs. The increase in the bolt load would equal the increase in bolt length, Δ, times the bolt-spring constant minus the gasket length decrease, Δ, times the gasket-spring constant. The magnitude of the added load would thus amount to-- Added load = $\Delta \times (k_B + k_G)$ and the increase in the load held by the bolt would equal $\Delta \times k_B$

The bolt-spring, k_B, constant equals $E_B A_B / L_B$ where E_B equals the modulus of elasticity of the bolt material, A_B equals the cross-sectional area of the bolt, and L_B equals the free length of the bolt.

$k_B = 2.1\times10^{11}$ x $0.00007854/0.150 = 11.00\text{x}10^7$ N/m

similarly for the gasket--with $A_G = 0.0088/12 = 0.000733$ m^2 per bolt

$k_G = 10.0\times10^8 \times 0.000733/0.0030 = 24.43\times10^7$ N/m

4.5 kN $= \Delta \times (11.00 + 24.43)\times10^7$

$\Delta = 4{,}500$ N/$(35.43\times10^7$ N/m$) = 127.0\times10^{-7}$ m

The increase in the load on each head bolt would equal- 127.0×10^{-7} x $11.0\times10^7 = 1{,}397$ N ■ which is considerably less than the load increase on the head of 4,500 N or 54/12 = 4,500 N/bolt.

It might be easier for some engineers who are completely unacquainted with the metric system to convert all the quantities to U.S. units, work the problem, and then convert the answer to metric units. There is an excellent table of conversion factors in the reference book "Fundamentals of Engineering Reference Handbook' supplied to all examinees. Using the foregoing problem as an example:

Bolt: Diam. = 0.3937 in. Area = 0.1217 in^2, L = 5.9055 in

$E = 2.1\times10^{11}/6{,}895 = 30.456\times10^6$ lb/in^2

Gasket: Area = 1.1367 in^2 per bolt L = 0.1181 in.

$E = 10^9/6{,}895 = 145{,}033$ psi

Then $k_B = 0.6276 \times 10_6$ lb/in and $k_G = 1.395 \times 10^6$ lb/in
The added head load per bolt = 4,500/4.448 = 1,017 lb
$\Delta = 1{,}017/(2.023 \times 10^6) = 503 \times 10^{-6}$ in. or 0.0005 in.
the increase in the bolt load 0.6267 × 503 = 315.2 lb
315.2 × 4.448 = 1,402 N ■

which is the same answer as obtained by the metric system calculation, considering slight errors in the conversion factors.

O-Ring Seal

It is interesting to note that there would have been no increase in the load on the vessel-head bolts if an O-ring seal had been used, see Fig. 1-4. This is a "hard" joint since the O-ring would not apply any load on the bolts. and thus the total load would equal only the original pre-load, at least until the pressure load exceeded the pre-load and the joint leaked.

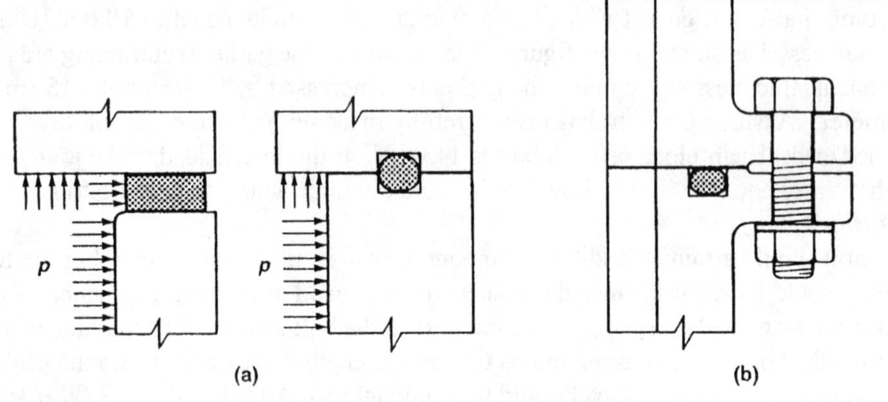

Figure 1-4

FATIGUE AND ENDURANCE LIMITS

Metal parts which are subjected to alternating loads will often fail when the applied stress is well below the strength of the material as determined by a static tensile test. Such failure usually starts at a point of discontinuity such as a change in cross-section or a local surface blemish where the localized stress is greater than the endurance strength of the material. Large forgings such as, for example, a large high-pressure triplex pump, can fail because of a rough machining mark at a point which is subjected to a stress which is well below the static strength of the material. A fatigue failure is a progressive failure and occurs over a period of time. Allowance is made for such reduced operating stress through the application of a "stress-concentration factor" which is applied to the endurance limit of the metal (usually steel) of which the part is made. In an extreme case--a high strength, heat-treated steel operating in sea water--the endurance limit may drop to as low as 11% of the ultimate strength.

The stress-concentration factor depends upon a number of different factors and includes the condition of the surface (roughness, polishing the surface of parts subjected to repeated stress variations has resulted in a considerably longer life, even showing an endurance strength approaching the static strength of a steel for a mirror-polished part), a size factor (sudden changes in size, like a change in diameter or a hole in the part), load factor (whether the loading reverses from tensile to compressive or varies entirely in tension), temperature factor, the medium in which the part is to be used (air, water, sea water), and other factors which may affect a particular component.

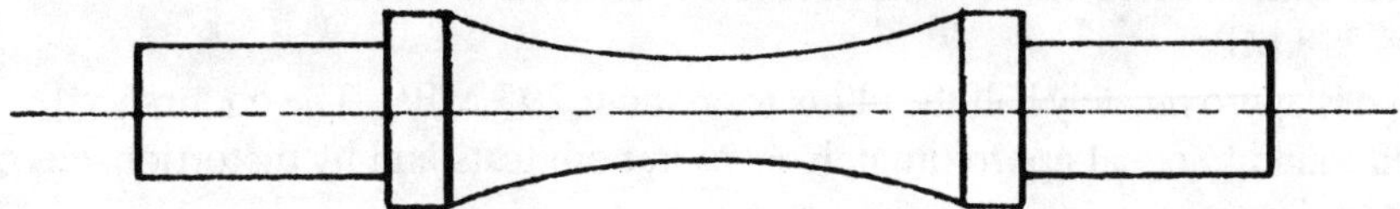

Figure 1-5

The endurance limit of a given metal is usually obtained by a rotating test wherein a sample of the metal which has been machined to a specific size and shape, see Fig. 1-5, has been rotated through many revolutions. Each revolution applies an equal tensile and compressive loading. When failure occurs the stress and number of cycles to failure are recorded and plotted on a graph. The endurance limit is that stress at which the specimen will withstand an infinite number of such cycles. The stress and number of cycles to failure are plotted on an S-N diagram, usually a semi-log diagram. The curve of stress vs failure usually flattens out for steels at about one to ten million cycles or more, but may not flatten out for aluminum even at many million cycles. The endurance limit may be defined as that stress at which a standard test specimen may withstand an infinite number of cycles of reversed stress. For ferritic carbon steels the endurance limit will usually average out to about 60% of the ultimate tensile strength, for pearlitic steels to about 40% of the UTS, and for plain carbon martensitic steels about 25%. For alloy martensitic alloy steels the ratio will average out to about 35%.

For design purposes this stress must be further modified (reduced) to take into account the stress-concentration effects previously noted.

One design method is that of the modified Goodman relation:

$S_a/S_e + S_m/S_{ult} = 1/N$ where N = safety factor

Where: S_a = the alternating tensile stress,

S_e = endurance limit

S_m = mean tensile stress, and

S_{ult} = ult. tensile strength

Another design method which is more conservative is with the Soderberg relation in which the value for the yield strength is used in place of the ultimate strength, giving:

$S_a/S_e + S_m/S_y = 1$ where S_y = yield strength

For shear or torsional stresses the Goodman relationship becomes-- $S_{vs}/S_{ns} + S_{ms}/S_{ults} = 1/N$

Where: S_{vs} is the variable shear stress,

S_{ns} the endurance limit in shear,

S_{ms} the mean shear stress,

S_{ults} the ultimate shear strength, and N the safety factor. The Soderberg relationship would be the same except the ultimate shear strength would be replaced by the shear yield stress S_{ys}.

EXAMPLE 1-11

A medium-carbon steel with S_u = 590 MPa and S_y =380 MPa is to be used for the design of a shaft which is to be subjected to varying torques ranging from +340 J to -115 J. What diameter should be specified if a safety factor of 2.0 is desired?

Solution:

Data for shearing stress for a specific steel may not be available, so in that case it must be estimated. The yield in shear is about 0.60 times yield in tension, so for this steel:

$S_{ys} = 0.60 \times 380 = 228$ MPa.

The endurance limit in tension is approximately half the ultimate or about 295 MPa. The endurance limit in torsion for steel has been found to equal approximately 60% in both tests and by distortion energy failure theory for polished specimens, so

$S_{ns}' = 0.60 \times 295 = 177$ MPa.

But this value must be further modified for surface effect and size effect. The surface effect for hot-rolled steel with a strength of 590 MPa is found from a graph to equal some 60%, and the size effect for a rod with a diameter between 13 and 50 mm equals some 85%. This gives a value for

$S_{ns} = 0.60 \times 0.85 \times 177 = 90.3$ MPa.

The mean torsional loading equals:

$T_m = [340 + (-115)]/2 = 112.5$ J

and the variable torque equals:

$T_v = [340 - (-115)]/2 = 227.5$ J

The corresponding stresses are then, from

$S = Tc/J'$ where $J'/c = \pi D^3/16$ so $S = 16T/(D\pi^3)$

$S_{ms} = 112.5 \times 16/\pi D^3 = 572.96/D^3$ Pa

$S_{vs} = 227.5 \times 16/\pi D^3 = 1{,}159.65/D^3$ Pa

Using the Soderberg relationship for a more conservative design gives--

$1/N = S_{ms}/S_{ys} + S_{vs}/S_{ns}$

$1/2 = 572.96/(228 \cdot D^3) + 1{,}159.65/(90.3 \cdot D^3)$

$D^3 = (5.03 + 26.68)/1{,}000{,}000$

$D = 0.0317$ m or 31.7 mm ■

EXAMPLE 1-12

Estimate the endurance limit of a 4.0 cm diameter cold- drawn steel bar for a 99 percent reliability, if it is made of a steel with an ultimate tensile strength of 400 MPa.

Solution:

The term "endurance limit" is generally considered to mean the bending endurance limit unless otherwise specified. From a table of recorded data we find that the size factor for this diameter equals 0.869. Also, for this size bar, the surface finish factor will equal 0.84. From another source it is found that the reliability factor for 99 percent reliability equals 0.814. Since we do not have any endurance test results for this particular steel we can estimate its endurance limit as 50 percent of the ultimate strength, or

$S_e' = 0.5 \times 400 = 200$ MPa

combining these factors gives--

$S_e = 0.869 \times 0.84 \times 0.814 \times 200 = 0.594 \times 200 = 119$ MPa ■

LOW CYCLE FATIGUE

There is another type of fatigue termed "Low Cycle Fatigue" in which failure occurs when a metal is stressed beyond its yield point in tension and then compressed beyond its yield point when the loading is reversed, thus causing the item to first to elongate plastically and then to shorten plastically, or vice versa.

EXAMPLE 1-13

One place where such failures have occurred is in the exhaust system of a diesel engine. If the exhaust pipe of a diesel engine is rigidly held at a point one meter from its attachment to the exhaust manifold, what would happen to the stainless steel pipe when the diesel was operated at full load for a full shift?

Solution:

The stainless steel exhaust pipe of the diesel near the exhaust manifold would probably become a bright red and its temperature would approach some 600 C or more. The coefficient of thermal expansion for stainless steel is 16.92×10^{-6}/oC. The pipe would tend to expand in length--

$\delta = 580 \times 16.92\times10^{-6} = 0.00981$ m/m

This would imply a compressive stress of--

$S = E\delta = 0.00981 \times 1.93\times10^{11} = 1{,}894$ MPa.

However, the yield strength of the stainless steel at room temperature is only 275.8 MPa, and this would decrease to as low as 130 MPa at the higher temperature, so the fixed-end piece of pipe would compress. The time at temperature is long enough for the pipe to "creep" to an equilibrium condition at an applied stress of 130 Mpa.

At the higher temperature--

$\delta = 130\times10^{6}/1.93\times10^{11} = 0.000674$ m/m

so the pipe would yield plastically--

0.00981 - 0.000674 = 0.00914 m/m or 9.14 mm over the one meter fixed length.

When the diesel engine was turned off, the pipe would cool and would attempt to shrink an amount equal to--

$\delta = 580 \times 16.92\times10^{-6} = 0.00981$ m/m, but this would mean a stress of--

$S = E\delta = 0.00981 \times 1.93\times10^{11} = 1{,}894$ MPa.

However, as previously noted, the yield strength of the stainless steel at room temperature is only 275.8 MPa, so the pipe would elongate plastically. This sequence of events would occur every time the diesel engine was operated, and after a number of such cycles (many many less than the endurance limit) the pipe would rupture. It is interesting to note that such a design was actually constructed and such a failure did occur. ■

SCREW THREADS

A power screw to raise a load is designed with a single square thread and a major diameter of 35.0 mm. The pitch is 5.0 mm. The bearing surface of the nut is 50 mm in diameter where it fits against the end the end of the structure to provide the resisting force. It is to be used to raise a load of 815 kg. It is assumed that the coefficient of friction for all surfaces equals 0.10. What would be the efficiency of the power screw, and what would be the stress in the screw?

Solution:

The pitch is 5 mm, so the width and depth of a square thread would equal 2.5 mm. The mean diameter would then equal OD minus the thread height or, $d_m = 35.0 - 2.5 = 32.25$ mm. When the screw is rotated, each turn will raise the mass 5.0 mm. The distance of the screw "wedge" would equal the circumference at the pitch diameter or $\pi \times d_m$ and

the thread lead angle $\lambda = \tan^{-1}(\text{pitch}/\pi\cdot d_m) = \tan^{-1}(5/32.25\pi) = \tan^{-1} 0.0494 \quad \lambda = 2.825°$

Probably the easiest way to analyze this problem is with the aid of the friction angle, μ, where the friction angle adds to, or subtracts from, the mechanical angle to give an equivalent frictionless system. The friction angle equals the $\tan^{-1}$ of the friction factor. A friction factor of 0.10 gives a friction angle equal to $\tan^{-1}0.10 = 5.711$o. The total equivalent frictionless wedge angle would equal $(\mu + \lambda)$ or 8.536o for raising the mass and $(\mu - \lambda) = 2.886$o for lowering the mass. It might be noted here that when the friction angle is greater that the lead angle, the screw will be statically stable, but if the lead angle is greater than the friction angle the force of the load will turn the screw and the load would lower unless the screw were held to keep it from turning. The torque required to turn the screw, considering only thread pitch and thread friction would equal--

$Tq_1 = (d_m/2)\cdot\tan(\lambda + \mu)\cdot$ Force

from $2\pi Tq = \pi d_m \cdot \tan(\lambda+\mu)\cdot F$

Torque would also be necessary to overcome the frictional resistance to turning due to the friction force acting between the bearing surface of the nut and the support structure.

$Tq_2 = F\cdot\mu\cdot(D_{od} + D_{id})/2 = F \times 0.1 \times (50 + 35)/2 = 4.25F$

the units in this case equal mm·N so to state it in joules or Nm

$Tq_2 = 0.00425\cdot F$ joules

The force, F, exerted by a load of 815 kg would equal--

$F = 815 \times 9.807 = 7993$ N so $Tq_2 = 33.970$ Nm

$Tq_1 = (32.25/2)\text{mm} \times \tan 8.536° \times 7993\text{N} = 19{,}345$ Nmm

or 19.345 Nm = 19.345 J

The total torque required to raise the load would equal--

Torque = $Tq_1 + Tq_2 = 19.345 + 33.970 = 53.315$ Nm = 53.315 J

Similarly, torque required to lower the load would equal--

$33.970 + 257.774 \times \tan 2.866° = 46.875$ J

The efficiency of the power screw would equal $F\cdot\text{pitch}/2\pi Tq$ so

$e = 7993\times0.005/(2\pi\cdot53.315) = 11.93\ \%$ ■

If a ball bearing thrust bearing were placed under the head of the bolt, assuming the same dimensions and a friction factor of 0.01, the torque Tq_2 would be reduced to 3.397 J and the total torque would equal 22.742 J. The efficiency would increase to--

$7993\times0.005/(2\pi\cdot22.742) = 27.97\ \%$

The shear area per thread would equal π times the root diameter (major diameter-2 times thread height) times the width of the thread--

Shear area = $\pi\cdot(0.035 - 0.005)\cdot0.0025 = 2.356\times10^{-4}\ \text{m}^2$

Stress for one thread of engagement--

$S_s = 7993/(2.356\times10^{-4}) = 33.93$ MPa

Which is a relatively low shear stress for steel, so, since there would be more than one thread of engagement, the screw thread is more than strong enough for the job.

BEAM DEFLECTION

A walkway is made from two steel tubes (pipes) 10 cm O.D. by 1.0 cm wall, spaced 60 cm apart and covered with wooden strips. It is used to span a 10 m ditch. If the mass of the wooden strips is ignored, what would be the maximum deflection if a man plus a wheelbarrow with a combined mass of 250 kg were to go across the bridge?

Solution:

The maximum deflection would occur when the man was at the center of the span. The deflection $\delta_{max} = P\ell^3/(48\cdot E\cdot I)$

This can be derived from the relationship for deflection for a simply supported beam with $b = \ell/2$ in the FE Handbook.

I for one tube equals $(\pi/64) \times (\text{O.D.}^4 - \text{I.D.}^4)$

$I = 0.04909 \times (1.000\cdot10^{-4} - 0.4096\cdot10^{-4})$

$I = 2.8983 \times 10^{-6}$ per tube so $I_{total} = 5.7966 \times 10^{-6}\ \text{m}^4$

$P = 250 \times 9.8066 = 2{,}452$ N

δ_{max} = 2,452 × 1,000/(48 × 2.1 × 10^{11} × 5.7966 × 10^{-6})
δ_{max} = 4.20 mm

What is the maximum stress in the span?
Solution:
This is a simply supported beam so the load on the support at each end would equal P/2 and the maximum moment would occur at the center of the span. $M_{max} = \ell \times P/4$
S = M·c/I = (10 × .0500 × 2,452/4)/(5.7966·10^{-6}) = 52.88 MPa ■

Problems

1-1.

A factor of safety is?

a) Yield stress divided by design stress.
b) Ultimate stress divided by design stress.
c) Ultimate stress divided by yield stress.
d) Maximum expected load divided by average load.

1-2.

A swing is to be designed for a public park. The support beam will be 3.00 m high, and the swing will be designed to hold a 100 kg person. What should be the strength of the lines holding a swing seat? Since this is a public park, it has to be assumed that the swing will be subjected to uses (or misuses) far beyond what any one might imagine, so a safety factor of five has been decided upon.

a) 12.2 kN
b) 13.9 kN
c) 14.7 kN
d) 15.6 kN

1-3.

A design requirement for a pogo stick is a spring that will compress 18 cm when a 100 kg person drops one meter. What is the required spring constant?

a) 65.7 kN/m
b) 68.8 kN/m
c) 71.4 kN/m
d) 75.2 kN/m

Solutions

1-1

The factor of safety is defined as the ultimate stress divided by the design stress. **Answer is b).**

1-2

The maximum load on the swing seat will occur at the bottom of the arc and will equal the gravitational force exerted by the swinger and the centrifugal force, F_c.

$F_c = m \cdot v^2/R$ Assume the swinger goes as high as the support beam, then $v = \sqrt{(2gh)}$ or

$v^2 = 2 \times 9.8066 \times 3.0 = 58.84\ m^2/s^2$ assuming the swing seat is at the level of the ground when at rest.

$F_c = 100 \times 58.84/3 = 1{,}961$ N

The total force down would equal

$1{,}961 + 100 \times 9.8066 = 2{,}942$ N

Assume that a swinger would hold onto only one side of the swing. The strength of each support line should equal $5 \times 2{,}942 = 14{,}710$ N.

Correct answer is c).

1-3

The energy to be absorbed equals--

$100 \times 9.8066 \times (1 + 0.18) = 1{,}157.18$ J

$1{,}157.18 = \frac{1}{2} \times k \times 0.18^2 = 71.431$ kN/m

The correct answer is c).

Chapter 2

Dynamic Systems, Vibration, Kinematics

The coverage of this subject is best described by a review of the chapter titled DYNAMICS in the "Fundamentals of Engineering (FE) Reference Handbook". This is the only reference permitted in the examination room and it behooves the exminee to become well acquainted with this reference. It will often be referred to in this chapter as the "Handbook".

Perhaps the best way to start a review of Dynamic Systems is to review Newton's laws, since the principles expounded by Newton govern most non-nuclear mechanical actions. The three laws can be expressed as follows:

1. When a body is at rest or moving with a constant speed in a straight line, the resultant of all the forces acting on the body is zero. (ΣForces = 0)

2. The rate of change of the momentum of a body is proportional to the force acting upon it and it is in the direction of the force. ($F = ma$)

3. Whenever one body exerts a force on another the second always exerts on the first a force which is equal in magnitude, but oppositely directed, and collinear. (Action and reaction are equal and opposite.)

LINEAR MOTION

(See Section KINEMATICS in the Handbook)

The basic relationships of motion should also be reviewed: Velocity, Acceleration, and Distance as discussed in the section titled KINEMATICS in the Handbook.

*Velocity*equals rate of change of distance with time, or $v = ds/dt$, which is a vector quantity, i.e. a velocity has both magnitude and direction. If the rate of change of position has only magnitude, but no direction, it is a scalar quantity--speed.

For straight-line motion and constant velocity, distance, s, equals $s_o + vt$ where s_o is the position at time zero.

Acceleration, a, is the rate of change of velocity, or $a = dv/dt = d^2s/dt^2$. If a constant acceleration acts on a particle then $ds/dt = v_o + at$ giving $ds = v_o dt + at\ dt$

Distance, Integration gives $s = s_o + v_o t + \frac{1}{2}at^2$ for straight-line motion.

Another important relationship is $v^2 = v_o^2 + 2a\Delta s$ where $\Delta s = (s - s_o)$, the distance traveled from time $t = 0$ to the instant when the new velocity is to be determined.

The four basic relations of linear (straight line) motion with constant acceleration can thus be stated as follows:

$v = v_o + at$ $\quad$ $v^2 = v_o^2 + 2v_o at + a^2t^2$

$v^2 = v_o^2 + 2a\Delta s$ $\quad$ $2as = 2a(v_o t + \frac{1}{2}at^2)$ $\quad$ $2as = 2v_o at + a^2t^2$ $\quad$ $v_{avg} = (v + v_o)/2$

$s = s_o + v_o t + \frac{1}{2}at^2$ $\quad$ or $s = v_{avg}t$ (For $S_o = 0$)

EXAMPLE 2-1

The speed of an automobile starting from rest increased in 18 seconds to 22 m/s. What was its final speed in km/hr? What was its average speed in km/hr during the 18 sec its speed was increasing?

Solution:

Assume a uniform rate of acceleration. Find also the rate of acceleration and the distance traveled.

From: $v = v_o + at$, $v = 22$ m/s, $v_o = 0$, $t = 18$ sec $a = 22/18 = 1.222$ m/sec² ■

the final speed in kilometers per hour equals-- 22 m/sec × (3,600 sec/hr)/1,000 = 79.2 km/hr

the average speed = (79.2 + 0)/2 = 39.6 km/hr

the distance traveled, $s = v_{avg} \times t$

$s = 39.6 \times 18/3{,}600 = 0.198$ km or $18 \times 22/2 = 198$ m

or, $s = \frac{1}{2}at^2 = \frac{1}{2} \times 1.222 \times 18^2 = 198$ m ■

Relative Velocity

Velocity is defined as the rate of change of distance (from some arbitrarily selected point) with time, or $v = ds/dt$. Thus a velocity possessed by an object must be in reference to some other object or reference system. That is, *all velocities are relative*. Usually we think of a velocity (i.e. any non-celestial velocity) in reference to the surface of the earth and tend to think of this as the velocity of an object. That can become confusing in some cases. Take, for example, the following problem–

EXAMPLE 2-2

A runaway railroad car is moving along a (frictionless) horizontal stretch of track at a constant velocity of 50 km/h. A locomotive starts out in pursuit and reaches a speed of 100 km/h when it is exactly 1.60 km behind the runaway car. The engineer (locomotive engineer, that is) starts to decelerate the engine at a constant rate so that when it touches the car to couple with it, the locomotive will be going at the same speed as the car. What should be the rate of deceleration and what additional distance would the runaway car travel before being caught?

Solution:

This problem can be worked by relating all velocities to the surface of the earth. If this approach is used the distance the car will travel can be denoted by D_c and the distance the locomotive will travel will thus equal D_l where $D_l = D_c + 1.6$ Then $100 \cdot t + \frac{1}{2}at^2 = 50 \cdot t + 1.6$ which gives: $50 \cdot t + \frac{1}{2}at^2 = 1.6$

With constant deceleration (negative acceleration) the average velocity of the locomotive would equal--

$v_{avg} = (100 + 50)/2 = 75$ km/h

The velocity of the car will remain constant $D_l = D_c + 1.6$ gives $75 \cdot t = 50 \cdot t + 1.6$

or $t = 0.064$ hr $s = 50 \times 0.064 + 1.6 = 100 \times 0.064 + \frac{1}{2} a \times 0.064^2$ which gives– $a = -781.25$ km/hr²

$D_c = 50 \times 0.064 = 3.2$ km ■

It is more straightforward and simpler to use relative velocities. The initial relative velocity of the locomotive in reference to the run-away car equals-- $v_r = 100 - 50 = 50 \text{ km/h} = v_o$

The final relative velocity equals zero. $v^2 = v_o^2 + 2 \cdot a \cdot s$ to obtain the final velocity

$0 = 50^2 + 2 \cdot a \cdot 1.6$ giving $a = -781.25$km/hr² $s = v_o \cdot t + \frac{1}{2} \cdot a \cdot t^2$ or $s = v_{avg} \times t$

$v_{avg} = 50/2 = 25$ km/hr $s = 1.6$ km $t = 0.064$ hr As before $D_c = 0.064 \times 50 = 3.2$ km ■

FLIGHT OF A PROJECTILE (See Section PROJECTILE MOTION in Handbook)

Another example of the use of equations of linear motion is afforded by a type of past examination problem which asked for details regarding the flight of a projectile.

EXAMPLE 2-3

A bullet leaves the muzzle of a gun at a velocity of 825 m/s at an angle of 30° with the horizontal, angle θ in the figure for projectile motion in the Reference Handbook. What is the maximum height to which the bullet will travel and how far will the bullet travel horizontally measured along the same elevation as the gun muzzle? Disregard air resistance.

Solution:

First determine the vertical and horizontal components of the velocity:

$v_v = 825 \sin 30° = 412.5$ m/s $\quad$ $v_h = 825 \cos 30° = 714.5$ m/s

The vertical component of the velocity of the bullet will be subjected to the acceleration of gravity, and if air resistance is disregarded, the time required for the bullet to reach its apogee can be determined from the fact that at the top of the bullet's flight the vertical velocity will equal zero.

$v = v_o + at$ or $0 = v_v - gt$ $\quad$ $t = v_v/g$ $\quad$ time to rise to apogee = 412.5/9.807 = 42.06 sec

The height to which the bullet will rise will equal– height = $v_{avg} \times t = (412.5/2) \times 42.06 = 8{,}675$ m

or, from $v^2 = v_o^2 + 2as$ which gives $h = v_v^2/2g$ $\quad$ $h = 412.5^2/(2\cdot9.807) = 8{,}675$ m $\quad$ $(a = -g)$

The distance the bullet will travel horizontally, the range, equals the time of flight times the horizontal velocity. The total time of flight will equal the time to rise to the apogee plus the time to fall back to earth, or twice the time to rise. $R = 2 \times 42.06 \text{ s} \times 714.5 \text{ m/s} = 60{,}103$ m ■

ROTARY MOTION (See Section KINETICS in the Handbook)

For plane circular motion the relationships corresponding to linear motion are--

θ, radians, corresponds to linear distance, s

$\omega = d\theta/dt$, radians per second, rate of rotation, corresponds to v, linear velocity

$\alpha = d\omega/dt$ or $d^2\theta/dt^2$, radians per second squared, corresponds to a, linear acceleration

$\omega = \omega_o + \alpha t$

In addition, the term for mass in linear motion is replaced by the moment of inertia "I" in rotational motion. This feature is discussed in detail later on in this chapter.

$\omega^2 = \omega_o^2 + 2\alpha\Delta\theta$

$\omega_{avg} = (\omega + \omega_o)/2$

$\theta = \theta_o + \omega_o t + ½\alpha t^2$

FORCE (See sections CONCEPT OF WEIGHT and CENTRIFUGAL FORCE in the Handbook.)

Newton's second law states $F = ma$ and the acceleration of gravity acts on a mass at the earth's surface in accordance with that relationship where $a = g = 9.8066$ m/s². The force of gravity exerted by a mass "m" = m × 9.8066 in newtons, thus the units of a newton N = kg·m/sec². Rearranging terms gives-- kg = N·sec²/m

There are also forces which are exerted due to rotation--the centripetal force (force acting inward toward the center of rotation) and the centrifugal force (force acting outward--centrifugal means "fleeing a center"). These forces are equal and opposite and the force is usually referred to as a "centrifugal force". An object moving in a circular path must have an acceleration toward the center of rotation or it would fly away from the circular path at a tangent as did the stone from David's sling as he pelted Goliath. Analysis shows that the acceleration of a particle in a circular path toward the center of revolution equals the square of the angular velocity times the radial distance or since force equals mass times acceleration, the centrifugal force, $F_c = \omega^2 r\cdot m$ or

$F_c = (v^2/r)\cdot m$

D'ALEMBERT PRINCIPLE

A sometimes useful relationship is termed the d'Alembert principle. This states that there is an "inertial force" which acts on any mass being accelerated, and equals ma. This derives from Newton's second law. Since F = ma, then F - ma = 0. The use of this term is illustrated in Example 2-4 below.

FRICTION (See Section FRICTION in the Handbook)

Friction is a passive, resisting force. It is equal to the coefficient of friction times the force *normal* to the surface and is independent of the area of contact. If the applied force parallel to the surface is greater than the force of friction, then the net force acting to cause motion will equal the applied parallel force minus the force of friction. If the friction force is greater than the component of the applied force parallel to the surface, no motion will occur and the system will be static. A frictional force is always a static resisting force, it is never an active force. The application of friction is illustrated in Example 2-4 below.

KINETIC ENERGY (See Section KINETIC ENERGY in the Handbook)

A mass in motion has energy, termed kinetic energy. This energy equals $\frac{1}{2}mv^2$ and is expressed in joules. $kg \times m^2/s^2 = N{\cdot}sec^2/m \times m^2/s^2 = N{\cdot}m = joules$
for example a 10 kg mass moving at 8 m/s possesses $\frac{1}{2} \times 10 \times 8^2 = 320$ J energy
$(N{\cdot}s^2/m \times m^2/s^2 = 320\ N{\cdot}m)$
Similarly a rotating mass contains rotational kinetic energy in the amount of $KE = \frac{1}{2}I\omega^2$ where I equals the moment of inertia of the mass about the center of rotation, and ω is the angular velocity in radians per second.

EXAMPLE 2-4

An example illustrating many of these principles is illustrated with the aid of Fig. 2-1. Assume a frictionless pulley and massless connecting line. For the system shown calculate (a) velocity of the 25 kg mass at impact and (b) the final distance of the 100 kg mass from the edge.

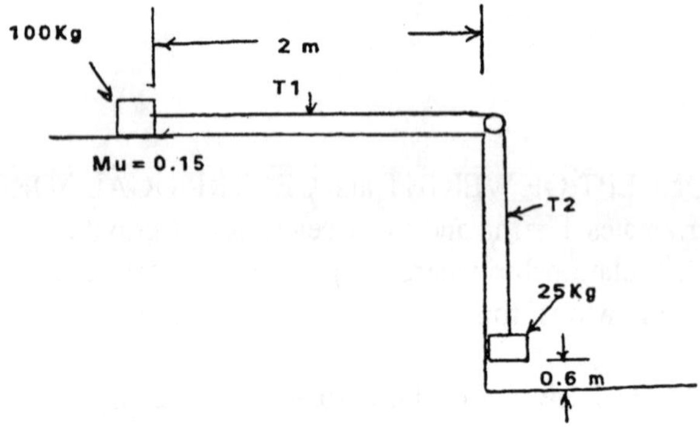

Figure 2-1

Solution:

The tension in the line equals $m_1 \times a + f$ where f equals the frictional force $= 100 \times 9.8066 \times 0.15 = 147.10$ N and $m_1{\cdot}a$ is the d'Alembert or resisting force.
so $T_1 = 100\,a + 147.10$ $T_2 = mg - ma = 25 \times 9.8066 - 25\,a$ Where mg is the force downward due to the force of gravity acting on the 25 kg mass, and $m{\cdot}a$ is the d'Alembert, or resisting force.
$T_1 = T_2$ $125\,a = 98.06$ N and $a = 0.7845\ m/s^2$
Velocity at impact $v^2 = v_o^2 + 2as = 0 + 2 \times 0.7845 \times 0.60$ $v = \sqrt{0.9414} = 0.9703$ m/s (a) ■
At the instant of impact the 100 kg mass would possess the same velocity, and the kinetic energy would equal: $KE = \frac{1}{2} \times 100 \times 0.9703^2 = 47.07$ J This energy would be expended by work done against friction, $s = 47.07/147.10 = 0.320$ m $2.00 - 0.320 = 1.680$ m from the edge (b). ■

EXAMPLE 2-5

What is the minimum force F to move the 50 kg mass over the level surface in Fig. 2-2 if the coefficient of friction is 0.4?

μ = coefficent of friction

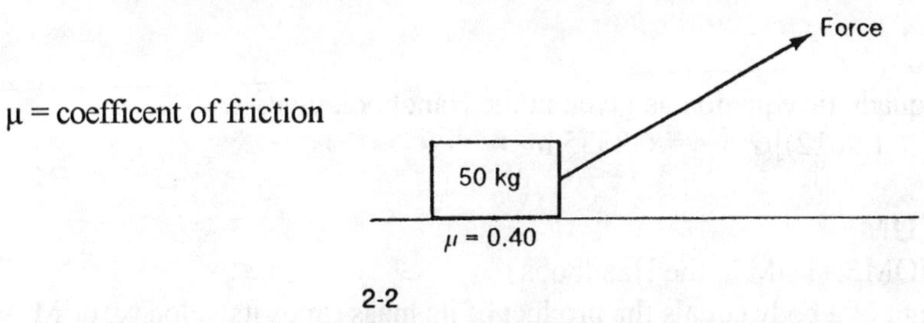

2-2
Figure 2-2

Solution:

Many engineers will merely glance at the problem and assume that the force required would equal $0.4 \times 50g$, but that is not correct because the force F has two components--a horizontal component which will cause the mass to slide, and a vertical component which will reduce the vertical force on the surface. Let the angle the force F makes with the horizontal equal α.

Then $(mg - F\sin\alpha)\mu = F\cos\alpha$

The force will be a minimum when $dF/d\alpha = 0$ expand and differentiate, giving--

$$-F\mu\cos\alpha\, d\alpha - \mu\sin\alpha\, dF = -F\sin\alpha\, d\alpha + \cos\alpha\, dF$$

combining terms gives--

$$dF/d\alpha = [F(\sin\alpha - \mu\cos\alpha)]/(\mu\sin\alpha + \cos\alpha) = 0$$

$F(\sin\alpha - \mu\cos\alpha) = 0$ and $\sin\alpha = 0.4\cdot\cos\alpha$

$\tan\alpha = 0.4$ $\quad \alpha = 21.80°$ $\quad 50g - 0.371F) \times 0.4 = 0.928\ F$

$F = 0.4 \times 50g/1.076 = 182.3\ N$ ■

POTENTIAL ENERGY (See Section POTENTIAL ENERGY in the Handbook)

Potential energy is energy contained by a body (mass) due to its position, or by a system due to its condition in which it possesses contained energy such as a spring, a fluid under pressure, an explosive, or similar arrangement. Two examples are a mass at an altitude and a compressed spring. If a mass is lowered or dropped a distance "s" the change in energy equals $\int Fds$ or $\int mgds$ and is measured in joules. Another form of potential energy is contained by a compressed spring. The force exerted by a spring = ks, where k = the spring constant, N/m. The potential energy equals $\int Fds$ or $\int ksds$ which equals, for $s_o = 0$, $\frac{1}{2}ks^2$, newton meters, or joules.

EXAMPLE 2-6

The velocity in the above problem (Example 2-4) can be checked by comparing the work done with the change in potential energy.

Solution:

ΔP.E. = ΔK.E. + work done = $25 \times 9.8066 \times 0.600 = 147.10\ J = \Delta K.E. + W_f$

W_f = work done against friction = $0.600 \times 147.10 = 88.26\ J$

so ΔK.E. = $147.10 - 88.26 = 58.84 = \frac{1}{2} \times 125 \times v^2$ $\quad$ m = 100+25 = 125 kg

$v = \sqrt{(58.84 \times 2/125)} = 0.9703$ m/s as before. ■

EXAMPLE 2-7

Determine how much a spring with a spring constant of 40 N/cm will compress if a mass of 50 kg is dropped onto it from a height of 2.00 m above the top of the spring.

Solution:

The loss of potential energy of the mass will equal: $50 \times 9.8066 \times (2.00 + s)$ where "s" equals the amount the spring compresses. This will equal the increase in potential energy in the spring:
$\Delta P.E. = \frac{1}{2} \times 4{,}000 \times s^2$ where "s" is in meters. $490.33 \times (2.00 + s) = 2{,}000\ s^2$
which reduces to:

$$s^2 - 0.2452s - 0.4903 = 0$$

Using the general solution for a quadratic equation as given in the Handbook gives--

$$s = [0.2452 \pm \sqrt{(0.0601 + 1.9612)}]/2 \quad s = 0.8335 \text{ m} \blacksquare$$

IMPULSE AND MOMENTUM

(See Section IMPULSE AND MOMENTUM in the Handbook)

Momentum, the momentum of a body equals the product of its mass times its velocity, or M = mv, assuming a constant mass. Impulse = $\int F dt$ or, since F = ma Impulse = $\int m(dv/dt)\cdot dt = \int m dv$, or impulse = change in momentum. If no external force acts on a system of bodies, the linear momentum is constant, this is expressed as the conservation of momentum. It should be noted that linear momentum is a vector quantity, since it is a vector quantity, v, times a scalar quantity, m. In addition, the angular momentum of a system of bodies about a fixed axis remains constant if no external moment acts about the axis. It is also true that the center of mass of a system remains constant if no external force acts on the system. These concepts can be applied to a number of different situations.

The ballistic pendulum used to measure the velocity of a bullet is one application.

EXAMPLE 2-8

If a bullet with a mass of one gram is fired at and becomes embedded in a block of mass two kg which is suspended by a cord 2.00 m long and causes the supporting cord to form an angle of 3.50° with the vertical. What is the velocity of the bullet at impact?

Solution:

The law of conservation of momentum states that the linear momentum of a system of bodies before impact equals the final linear momentum after impact if there is no resultant external force acting on the system, or $m_1 v_1 = (m_1 + m_2)\cdot v_2$ the pendulum block will rise– $2 \times (1 - \cos 3.5°) = 0.003730$ m. The change in potential energy equals the change in kinetic energy.
So $v = \sqrt{2gh}$, or $v = \sqrt{(19.6132 \times 0.003730)} = 0.2705$ m/s
$v_1 = 2.001 \times 0.2705/0.001 = 541.2$ m/s $\blacksquare$

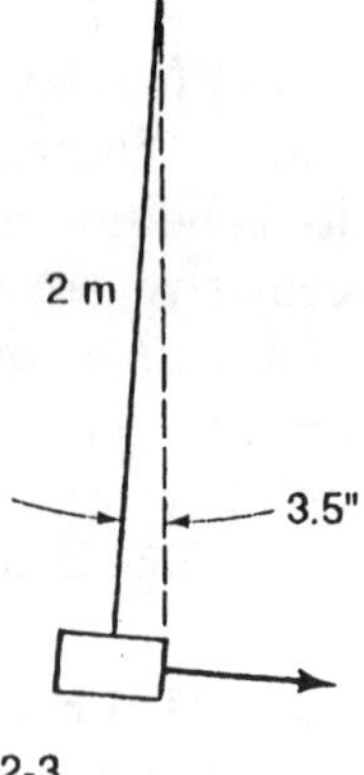

2-3

Figure 2-3

COEFFICIENT OF RESTITUTION

Many impacts are not plastic and in many cases one of the impacting bodies will "bounce". If a ball rebounds upward after being dropped onto a fixed plate, the mass of the plate being much greater than that of the ball, then the velocity after impact will be less than the velocity at impact, and the coefficient of restitution, e, will equal the negative of the ratio of the two velocities.

$$e = -v_2/v_1 = \sqrt{(h_2/v_1)} \quad (v_2 \text{ is opposite in sense to } v_1)$$

The coefficient of restitution for two bodies whose masses do not differ widely is equal to the ratio of the velocity of separation, $v_1 - v_2$, divided by the velocity of approach, $u_2 - u_1$, or:

$e = (v_1 - v_2)/(u_2 - u_1)$, as noted in the Handbook.

EXAMPLE 2-9

A ball with a mass of 4.0 kg traveling with a velocity of 6 m/s overtakes and strikes a 6.0 kg ball traveling in the same direction with a velocity of 2 m/s. If the coefficient of restitution equals 0.75, what would be the velocities of the two balls after the impact?

Solution:

$e = 0.75 = -(v_2 - u_2)/(6 - 2)$ which gives $u_2 - v_2 = 3$ but $m_1v_1 + m_2u_1 = m_1v_2 + m_2u_2$ so

$4{\cdot}6 + 6{\cdot}2 = 4{\cdot}v_2 + 6{\cdot}u_2$ and $9 = v_2 + 1.50{\cdot}u_2$

Combining the two equations gives--

$v_2 = 1.80$ m/s $u_2 = 4.80$ m/s

Another example of the principles of Impulse and Momentum is afforded by the following problem:

EXAMPLE 2-10

Two barges, one of total mass 10,000 kg and another of 20,000 kg are connected by a cable in still water. The two barges are initially connected by a cable 30 m long and are at rest. If the cable is drawn in by a winch on the larger barge until the distance separating the two barges is 15 m, what would be the distance moved by the 10,000 kg barge? Assume friction is negligible.

Solution:

The key point in this problem is that the center of gravity of a system cannot change if no external force is applied to the system.

The impulse will be the same on each of the two barges and the center of gravity of the system will remain the same since no external force acts on the system. Initially the CG will be a distance of– 30/3 = 10 m from the larger barge and 20 m from the smaller barge.

$10{,}000g \times s = 20{,}000g \times (30 - s)$ $s = 20$ m so, initially the CG will lie 2/3 the distance between the small and the large barge. The location of the CG will not change so the small barge will move 10 m and the larger barge will move 5 m. ■

Another illustration of the principles of impulse and momentum is afforded by a golf ball which is struck by a club.

EXAMPLE 2-11

High-speed photography shows that the club is in contact with the ball for approximately one-half-thousandth of a second and the velocity of the ball was 75 m/s immediately after being struck. What force is exerted by the club if the mass of the ball is 47 grams?

Solution:

Assume the force is constant during the period of impact, so the impulse will equal 0.0005F. The change in momentum of the golf ball will equal Δmv, which gives--

$\Delta mv = 47 \times 75 = 3{,}525$ gm·m/s or 3.525 kg·m/s

Since the change in momentum is equal to the impulse

$0.0005F = 3.525$ $F = 7{,}050$ N ■ kg·m/s = F·s or F = kg·m/s² = Newtons

It should be noted that the concept of conservation of momentum does not mean conservation of energy. This is demonstrated by the following problem.

EXAMPLE 2-12

A plastic body with a mass of 45 kg is moving with a velocity of 6 m/sec and overtakes another plastic body with a mass of 70 kg, moving in the same direction with a velocity of 4 m/sec. What is the final velocity of the combined masses and what is the loss in kinetic energy?

Solution:

From the principle of conservation of momentum we have $m_1v_1 + m_2v_2 = (m_1 + m_2)v_r$
The resulting velocity equals–

$v_r = (45{\cdot}6 + 70{\cdot}4)/(45 + 70) = 4.783$ m/sec ■

KE at start = $\frac{1}{2}{\cdot}45{\cdot}6^2 + \frac{1}{2}{\cdot}70{\cdot}4^2 = 1{,}370$ J

Final KE = $\frac{1}{2}{\cdot}115{\cdot}4.783^2 = 1{,}315$ J ■

The loss of kinetic energy equals 55 J or 4.0% which amount of lost mechanical energy is converted to thermal energy which results in heating up the two bodies.

MOMENT OF INERTIA

(See Section MOMENT OF INERTIA in Chapter titled STATICS in the Handbook, also review table listing Area & Centroid, Moment of Inertia, (Radius of Gyration)², and Product of Inertia for a number of different figures in that chapter}.

Note that $J = \int r^2 dA$ (here J is the polar moment of inertia), and $J = k^2A$ where k = radius of gyration. So in the table k^2 is given (the symbol, r, is used in the Handbook rather than k). The values given in the table are for plane figures. To obtain the values of "I" for a solid it is necessary to introduce the mass of the solid. For example the mass moment of inertia for a cylinder which is rotating about an axis through its center of gravity, or the center of the circle shown, equals $\frac{1}{2}Mr^2$. The value of J for the plane polar moment of inertia equals $\int r^2 dA$, whereas the mass polar moment of inertia J_M equals $\int r^2 dM$, but Mass = density × volume, or $\rho \times L \times A$.

Then $dM = \rho L dA$ and $J_M = \rho L \pi r^4/2 = \rho{\cdot}V{\cdot}r^2/2$ or $J_M = \frac{1}{2}Mr^2$ It might also be noted that the polar moment of inertia equals the sum of the moments of inertia about any two axes at right angles to each other in the plane of the area and intersecting at the pole. It is interesting to note how this concept simplifies the calculation of the transverse moment of inertia of a circular section about a transverse axis through its center in the plane of the surface. The solution of the relationship $\int x^2 dA$ about an axis in the plane of the disk is quite complex. It is relatively easy to calculate the polar moment of inertia of the disk, and half of the polar moment of inertia equals the transverse moment of inertia since the moments of inertia of a circular disk at right angles are the same. Then the transverse moment of inertia equals

$\frac{1}{2}J = \frac{1}{2}\int r^2 dA = \frac{1}{2}\int r^2{\cdot}2\pi r dr$

since $dA = 2\pi r dr$ which, integrated from 0 to R gives

$J = \pi R^4/2$ or $AR^2/2$ and I (transverse) = $\frac{1}{4}AR^2$

The corresponding relationships for linear and rotary motion are described above, however there are other relationships for rotary motion which are also important, The quantity of mass in a rotational problem is replaced by the Polar Moment of Inertia,

where $I = \int r^2 dm$ ($J = \frac{1}{2}MR^2$)

which gives the relationship of $T = I\alpha$ corresponding to $F = ma$,

where T is the torque in Nm.

In addition, K.E. = $\frac{1}{2}I\omega^2$, for rotary motion which corresponds to the linear K.E. = $\frac{1}{2}mv^2$. N.B. It is customary to refer to the moment of inertia in rotary motion as "I", rather than "J", as, for example in $T = I\alpha$.

Another important relationship in rotary motion is the concept of centrifugal force, F_c, or force outward on an object due to its rotation about some point, e.g a satellite rotating about the earth.

$F_c = \omega^2 rm$ or $F_c = (v^2/r)m$

where ω is the rotational velocity in radians per second and v is the linear velocity in m/s, since $v = \omega r$. Note that "r", the radius, is the distance from the center of mass of the object to the center of rotation of the system. In the case of an earth's satellite the center of rotation would be the center of the earth, and the distance "r" would be the distance from the center of the earth to the center of mass of the satellite.

As noted above, the mass rotational or polar moment of inertia,

$I = \int r^2 dm$ and $I = k^2 m$

where "k" is the radius of gyration,

or the distance from the axis of interest at which all of the mass can be considered concentrated as, for example a hoop with negligible thickness. It should be noted that, in general, the mass of a body cannot be considered as acting at its center of mass for the purpose of calculating the moment of inertia. This can easily be shown by calculating the moment of inertia of a uniform slender rod about an axis through one end and perpendicular to the rod, see Fig. 2-4.

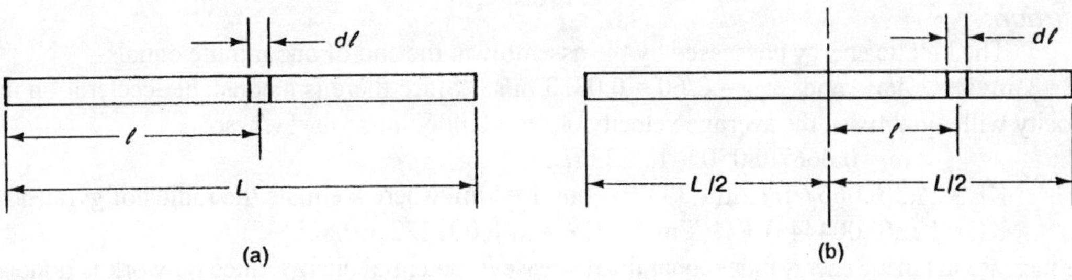

2.4

Figure 2-4

The mass of the rod will equal ρAL where ρ is the density of the rod material in kg per unit volume and A is the cross-sectional area and L is the length. The dm of the relationship for $I = \int r^2 dm$ will then equal $\rho A d\ell$ giving

$I = \int \ell^2 \rho A \, d\ell$ for $\ell = 0$ to $\ell = L$

I then equals $\frac{1}{3}\rho AL^3$

and since the mass, M, of the rod $M = \rho AL$ $I = \frac{1}{3}ML^2$ rather than $ML^2/4$ $[M(L/2)^2 = ML^2/4]$

The distance from the end of the rod to the center of mass equals L/2, but the length of the radius of gyration from the end of the rod equals $L/\sqrt{3}$.

Similarly, the moment of inertia of a slender, uniform rod about an axis through its center and perpendicular to the rod will equal $\int \ell^2 dm$ from -L/2 to +L/2 giving--

$I = (1/12)ML^2$

and the moment of inertia of a flat, rectangular surface about an axis through its center lying in the plane of the surface will equal, see Fig 2-5:

$I = \int x^2 b \, dx$ from $x = h/2$ to $x = -h/2$

which gives-- $I = bh^3/12$

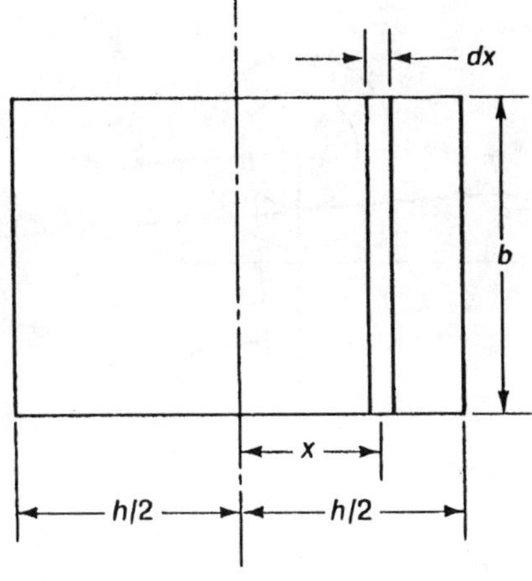

2-5

Figure 2-5

EXAMPLE 2-13

An armature is keyed to a shaft 10 cm in diameter. It rests transversely on two parallel steel rails inclined to the horizontal with a slope of 1:12. See Fig. 2-6. The shaft rolls, without slipping and without rolling friction, a distance of two meters along the rails from rest in one minute. What is the radius of gyration of the armature and its shaft about the assembly's longitudinal axis?

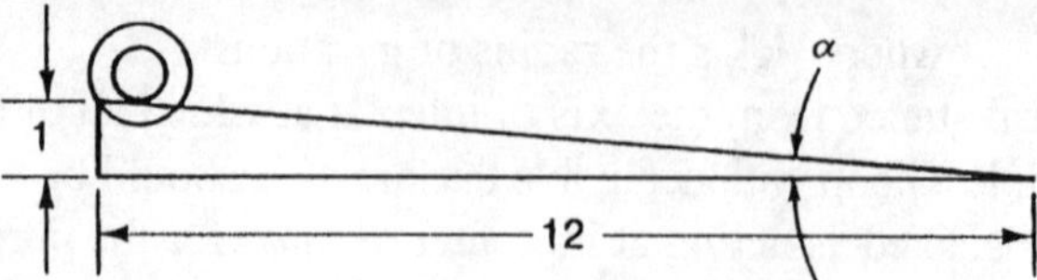

Figure 2-6

Solution:

The kinetic energy possessed by the assembly at the end of one minute equals–

$KE = ½mv^2 + ½I\omega^2$ and $v_{avg} = 2/60 = 0.0333$ m/s Since there is a constant acceleration the final velocity will equal twice the average velocity or $v = 0.0667$ m/s $\omega = v/r$ so:

$$\omega = 0.0667/0.050 = 1.3333 \text{ rad/s}$$

$KE = ½m(0.0667^2) + ½I(1.3333^2)$ but $I = k^2m$ where k equals the radius of gyration. So--

$KE = ½m(0.004445) + ½(k^2m)1.7779 = m(0.002222 + 0.8888k^2)$

The increase in kinetic energy must equal the decrease in potential energy since no work is done against friction. The angle of the slope, α, equals $\tan^{-1}(1/12) = 4.76°$

$\Delta PE = mg \sin\alpha \times 2.0 = 9.807 \times 0.08305 \times 2 \times m = 1.6289 \times m$

ΔP.E. equals ΔK.E. so

$0.002222 + 0.8888\,k^2 = 1.6289$ giving $k = \sqrt{1.8304} = 1.353$ meters ■

EXAMPLE 2-14

The same result can be obtained by determining the acceleration of the armature assembly with the relationships that $Tq = I\alpha$ and $I = mk^2$ See Fig. 2-7 $s = v_ot + ½at^2$ where $v_o = 0$ and $t = 60$ sec $2 = ½\cdot a\cdot 60^2$ $a = 0.00111$ m/sec²

Solution:

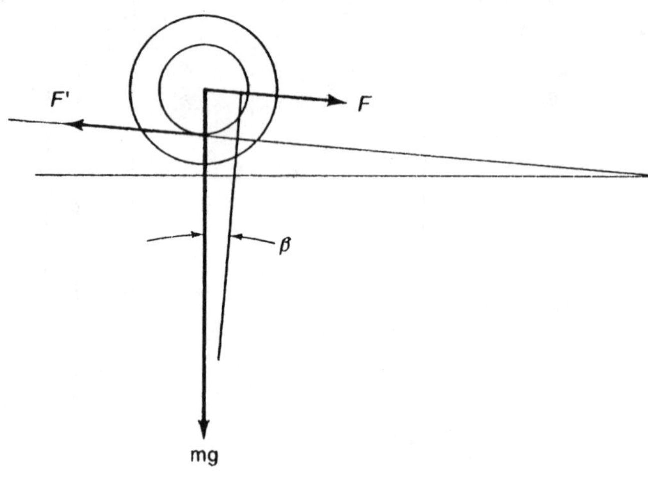

2-7

Figure 2-7

The force acting on the assembly to cause it to accelerate down the incline will equal its mass times the gravitational constant times the sine of the slope angle, and since there is no slippage, the resisting force applied by the rails will be equal and opposite to the applied force, F, giving $F = F'$

$F = mg \sin\alpha$

$= m\cdot 9.807\cdot 0.08305 = 0.8145\cdot m$

$Tq = I\alpha = I(a/r)$

$Tq = F' \times r = F \times r$

so $0.8145\cdot m\cdot r = k^2\cdot m\cdot(a/r)$ which gives– $k^2 = 0.8145\cdot r^2/a$

$= 0.8145\cdot 0.050^2/0.00111 = 1.8345$

which gives $k = 1.354$ meters as before. ■

This example brings up an important point about the summation of the forces acting on the armature, or any rotating body subjected to rotational movement along a surface. In the illustrated case the frictional force, F', was just equal to the applied force, F, and was balanced by it. There was no excess force to cause a sliding, linear acceleration and, α, the rotational acceleration equaled the linear acceleration divided by the rolling radius. If, however, F had exceeded F', the armature would have slipped as well as rolled and there would have been a linear acceleration in addition to the angular acceleration. The effect of the greater force can be illustrated by a past exam problem as follows:

EXAMPLE 2-15

A cylinder with a mass of 30 kg and a radius of 1.25 m is pushed by a moving bulldozer, see Fig. 2-8. The coefficient of friction between the blade of the bulldozer and the cylinder and between the cylinder and the surface it rests upon is 0.20. Determine the minimum acceleration of the bulldozer for the cylinder to slide along the surface without rotating, and the maximum acceleration of the bulldozer if the cylinder is to roll without slipping.

Figure 2-8

Figure 2-9

Solution:

The maximum frictional force between the cylinder and the surface it rests upon will equal F'= 0.20·mg. The net torque acting on the cylinder will equal, see Fig 2-9--
$F' \times R - F_B \times R$ where F_B is the frictional force between the blade of the bulldozer and the cylinder at the point of contact. The blade is assumed to be flat and it contacts the cylinder at a point at the same elevation as the center of the cylinder. From the equation for the torque acting on the cylinder it can be seen that if F_B is greater than F' the cylinder will not roll. Similarly if F' is greater than F_B the cylinder will roll rather than slide. The critical acceleration where the cylinder will either slide or roll with a change in the acceleration of the bulldozer will occur when $F' = F_B$ At this point the force exerted by the blade of the bulldozer on the cylinder must equal the same as the gravitational force exerted by the cylinder on the surface it rests on. The inertial, or d'Alembert, force must then equal mg or the critical rate of acceleration of the bulldozer equals the acceleration of gravity. At any lower rate of acceleration the cylinder will roll, and at any greater rate of acceleration the cylinder will slide.

TRANSFER OF MOMENT OF INERTIA

As discussed in the Handbook, the moment of inertia of a body or an area about another point or axis parallel to the axis through the area's centroidal axis can be determined by means of the parallel axis theorem which states the I_p, the moment of inertia about a point "p" equals the moment of inertia of that body about its centroidal axis plus its mass (or area) times the square of the distance from the centroidal axis. This principle is of more importance in the determination of stresses in the study of mechanics of materials than it is in this chapter, and it will be treated in more detail there.

EXAMPLE 2-16

The application of many of the principles of inertia, gravity, and rotational forces can be illustrated by estimating how high a satellite must be located to rotate around the earth once each day.

The force out would equal the centrifugal force, this would be balanced against the gravitational force exerted by the earth. The force of gravity existing between two objects equals a constant times the product of the two masses divided by the square of the distance between the two centers of mass: $F_g = K(m_1m_2/s^2)$. The force of gravity on a mass at the earth's surface equals 9.8066 times the mass. The radius of the earth is approximately 6,437 km. F = mg or
$F = (m_1 \times 9.8066)$ Newtons $F = K(m_1m_e/6{,}437{,}000^2)$ so $K = 4.063 \times 10^{14}/m_e$
where m_e = mass of the earth in kg $\omega = 2\pi/(24 \times 3{,}600) = 0.00007272$ rad/s
The centrifugal force, $f_c = 5.288 \times 10^{-9}$ s·m which equals the gravitational force exerted by the earth.
$5.288 \times 10^{-9} \times s{\cdot}m = [4.063 \times 10^{14}/m_e] \times mm_e/s^2$ s = distance between the CGs of the earth and the satellite. $s^3 = 0.7683 \times 10^{23}$ or s = 42,506,000 meters or 42,506 km, so the height of the satellite must be 42,506 - 6,437 = 36,089 km above the surface of the earth (6,437 km = approx radius of earth).

EXAMPLE 2-17

Another example is illustrated in Fig. 2-10.
A marble of radius r is allowed to roll down an incline at the bottom of which it enters a loop-the-loop of radius R. How high above the bottom of the loop-the-loop must the top of the incline be, distance h, for the marble to remain in contact with the loop at the top? I for a sphere equals $2/5mr^2$. Assume the marble rolls without slipping and no energy is lost to friction.

Solution:

At the top of the loop the energy contained by the rolling marble will equal its translational energy due to its velocity plus it rotational energy. The centrifugal force must equal the gravitational force $\omega^2 r_e m = mg = (v^2/r_e) \times m$
where $r_e = R - r$ (R = radius of loop and r = radius of marble) then $v_T = \sqrt{[g \times (R - r)]}$
to determine the velocity at the bottom, the energy will increase an amount equal to the change in potential energy

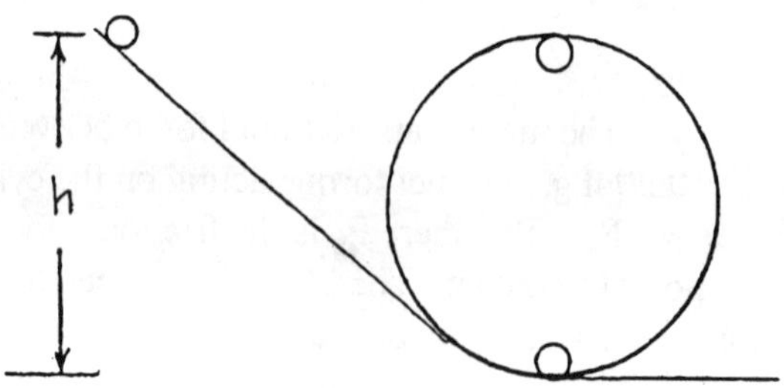

Figure 2-10

ΔK.E. = ΔP.E.

ΔP.E. = $mg\Delta h = mg \times 2(R - r) = ½mv^2 + ½I\omega^2$

at top K.E. = $½mv_T^2 + ½ \times (2/5 \times mr^2)v_T^2/r^2$
where ω=v/r
Giving an amount of energy at the top of loop

K.E. = $0.700\ m{\cdot}v_T^2 = 0.700m \times g(R - r)$

at bottom of loop, energy equals:

K.E. at top + $mg \times 2(R - r) = 2.70 \times mg \times (R - r)$

$mgh = 2.70\ mg \times (R - r)$ or $h = 2.70 \times (R - r)$ ■

EXAMPLE 2-18

A homogeneous cylinder of radius **r** rolls down a plane inclined at an angle of 30° with the horizontal. (a) What is the rate of acceleration of the center, and (b) what is the minimum coefficient of friction, μ, to prevent the cylinder from slipping? See Fig. 2-11.

Solution:

The force acting at the CG of the cylinder parallel to the plane equals mg sin 30o = 0.500 mg. Torque = Iα The torque equals 0.500 mg × r. The moment of inertia about the point of contact, I_c, can be calculated by means of the parallel axis theorem.

$I_c = \frac{1}{2}mr^2 + mr^2 = 1.5\ mr^2$

torque = 0.500 mg × r = 1.5 mr² α α = a/r a = g/3

the answer to part (a) a = 3.27 m/s² ■

For part (b) ΣF on cylinder 0.500 mg - μmg × 0.866 = ma = mg/3

giving

$\mu = 0.1667/0.866 = 0.1925$ (b) ■

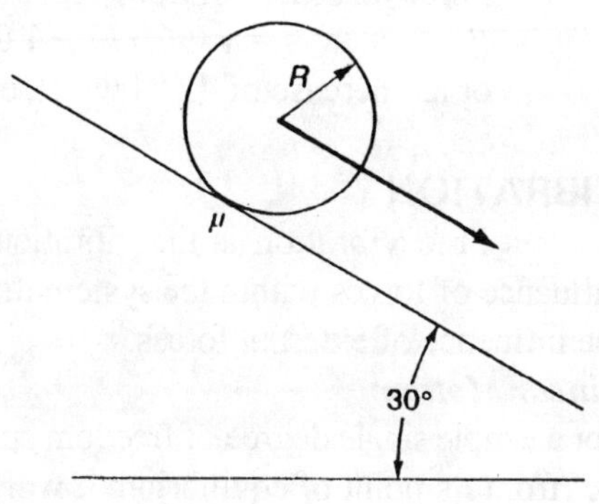

Figure 2-11

BANKING OF CURVES

Another type of problem utilizing the principles of centrifugal (or centripetal) force is the calculation of the superelevation (slope of roadway) required in the construction of a highway for an assumed particular speed of automobile.

EXAMPLE 2-19

What superelevation, β, is required on a highway where the automobile design speed is taken as 100 km/hr and the curve of the radius is to be 765 m, if a driver is to exert no sideward force on car seat? If the mass of the driver equals 90 kg, how much force will he exert perpendicular to the seat?

Solution:

The forces acting on the car will equal the force of gravity down and the centrifugal force acting outward. The two forces will form the legs of a right triangle whose angle will equal the required angle of superelevation. See Fig. 2-12.

The acceleration acting down would equal--

F = mass·g = 9.807·mass

The centrifugal force acting at a right angle to the force of gravity would equal

$\omega^2 \cdot r \cdot m$ or $(v^2/r) \cdot m$

$(v^2/r) \cdot \text{mass} = (100{,}000/3600)^2/765$

$= (1.008\ m/s^2) \cdot \text{mass}$

The required angle of superelevation would equal–

$\beta = \tan^{-1} 1.008/9.807 = 5.87°$

Combining terms and reducing the relationship to its simplest form gives $\beta = \tan^{-1} v^2/(gr)$ which is the relationship given in the Handbook. $\tan^{-1}(1.008/9.807) = 5.87°$ as before ■

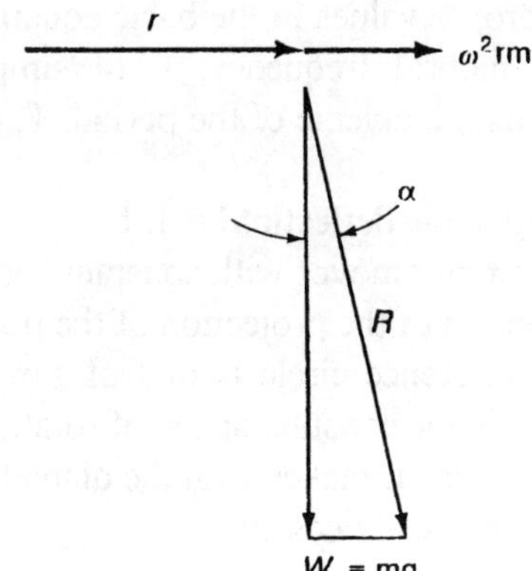

2-12

Figure 2-12

EXAMPLE 2-20

The force the driver would exert on the seat would equal the square root of the sum of the squares of the two components of the acceleration times his mass.

Solution:

Force on seat = $\sqrt{(9.807^2 + 1.008^2)} \times$ mass

$= \sqrt{(96.177 + 1.016)} \times 90 = 9.859 \times 90 = 887.28$ N ■

or an increase of 4.7 N which equals 0.5%

VIBRATION

Free vibration is the vibration that takes place when an elastic system vibrates under the influence of forces within the system itself. Forced vibration is the vibration which takes place due to the influence of external forces.

Linear Motion

For a simple single degree of freedom such as is shown in Fig. 2-13, if the mass is displaced a distance "x" from its point of equilibrium it will be acted upon by a restoring force equal to -kx. The net force acting on the mass equals

$F = mg - k(x + \Delta x) = ma$ but $mg = k\Delta$ so $F = -kx$

or $-kx = ma = m\, d^2x/dt^2$

which is a second order differential equation, the general solution to which is:

$x = A \sin \omega t + B \cos \omega t$

SIMPLE HARMONIC MOTION

Simple harmonic motion is defined as the motion of a point in a straight line such that the acceleration of the point is proportional to the distance of the point from its equilibrium position, or–

$a = -kx = -d^2x/dt^2$.

An example of simple harmonic motion is shown by the suspended mass shown in Fig 2-13 and described above.

Solution:

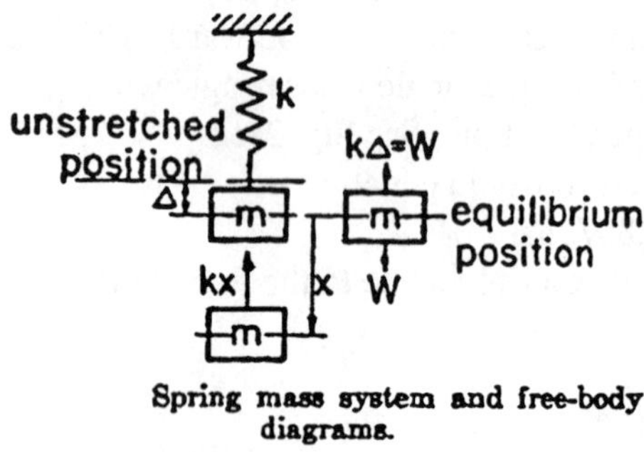

Spring mass system and free-body diagrams.

Figure 2-13

Substitution of the proper values in the basic equation gives the undamped natural frequency, f, of simple harmonic motion equals the inverse of the period, T, or $f = 1/T$

Then: $f = 1/2\pi \times \sqrt{[g/\text{static deflection}]} = 1/T$

It can be shown that if a point moves with constant speed in a circular path the motion of the projection of the point on a diameter of the reference circle is that of simple harmonic motion. If ω is the constant speed of rotation, and θ is the angle the radius, r, makes with the diameter, then $\theta = \omega \cdot t$ and $x = r \cdot \cos\theta = r \cdot \cos \omega t$

See Fig. 2-14.

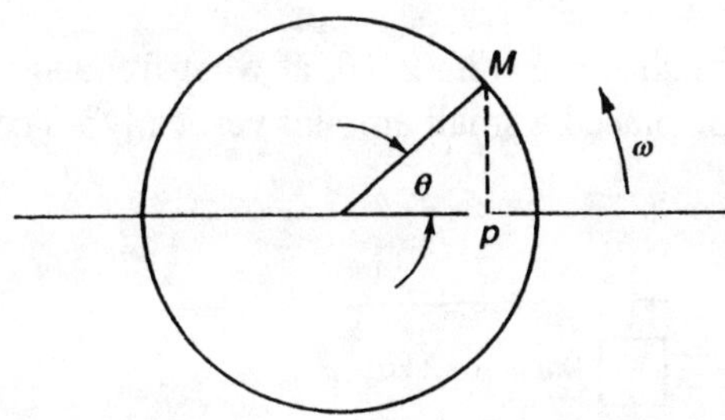

Figure 2-14

The velocity of the point on the reference diameter, i.e. of the object subjected to simple harmonic motion, $v_p = dx/dt = -\omega r \sin \omega t = -\omega y$ and the acceleration of the point (or body) $a_p = dv_p/dt = -\omega^2 r \cos \omega t = -\omega^2 x$ which is the equation for simple harmonic motion. The time for the point P to make one complete oscillation is the same as the time for the point M to make one complete revolution, or $T = 2\pi/\omega$, and the frequency equals $1/T = \omega/2\pi$. ■

EXAMPLE 2-21

A 50 g mass is resting on a spring with a spring constant of $k = 100$ N/m, what is the natural frequency of the system? See Fig. 2-15

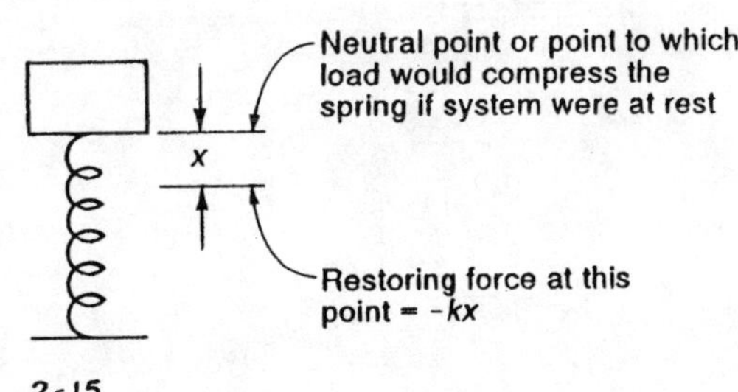

Figure 2-15

Solution:

The force, F, exerted on spring equals--
$F = 0.050 \times 9.8066 = 0.4903$ N
Deflection = 0.4903/100 = 0.004903 m
And frequency
$f = 1/2\pi \times \sqrt{(9.8066/0.004903)} = 7.12$ c.p.s. ■

EXAMPLE 2-22

The base of a one-kg instrument is set on four rubber mounts, each of which is rated at 5.0 mm deflection per kg of load. What is the natural frequency of vibration?

Solution:

The deflection of the four rubber mounts together is: 0.005/4 = 0.00125 m
$f = (1/2\pi) \times \sqrt{(9.8066/0.00125)} = 14.10$ c.p.s. ■

EXAMPLE 2-23

A truck body lowers 15 cm when a load of 12 metric tons is placed on it. What is the frequency of vibration of the loaded truck if the total spring-borne mass is 17 metric tons?

Solution:

Frequency $= 1/T = \omega/2\pi$, and the acceleration of the mass $a = -\omega^2 x$ but $F = ma$, and the force acting on the mass equals $-kx$, so $-kx = -\omega^2 mx$ and $\omega = \sqrt{(k/m)}$
$f = \omega/2\pi = (1/2\pi)\sqrt{(k/m)}$ or static deflection = (17/12) x 0.15 = 0.2125 m
The spring constant equals 12,000·9.807/0.15 = 784,560 N/m
Combining terms gives--
$f = (1/2\pi)\sqrt{[(9.807/0.2125)]} = 1.081$/sec
which gives--
$f = \text{constant} \times \sqrt{[(N/m)(m/sec^2)/N]}$ = oscillations/sec ■
Note that the restoring force acting on the truck body equals $-kx$ where x = the distance the spring is compressed past the neutral point.

EXAMPLE 2-24

If the lever is assumed to be mass-less in the system shown in Fig. 2-16, at what frequency would the mass of 12 kg move up and down after it had been displaced a small amount vertically? The spring constant, k, equals 75 N/m.

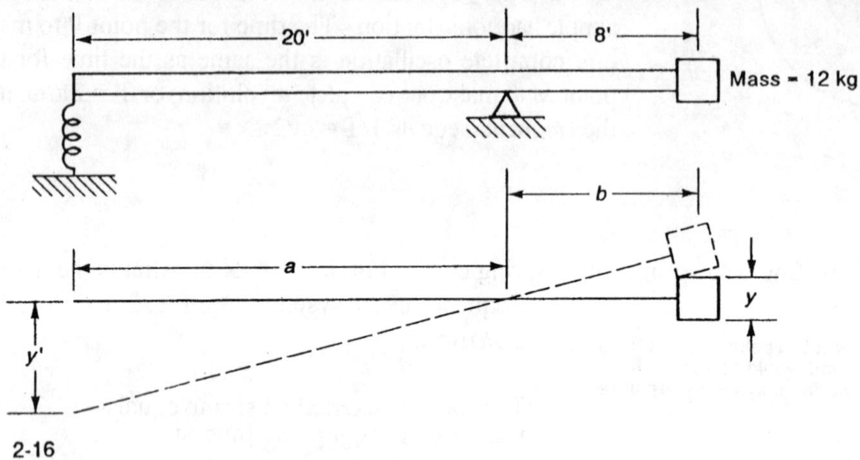

Figure 2-16

Solution:

This is an example of undamped free vibration with a single degree of freedom. For small displacements the system will vibrate in conformance with the relationship for simple harmonic motion in the same way as the suspended mass shown in Fig. 2-13. Thus the motion will satisfy the differential equation– $d^2x/dt^2 + (k/m)\cdot x = 0$ as given in the Handbook, or in this case
$d^2y/dt^2 + (k/m)\cdot y$ since the displacement is vertical.

The general solution of this equation gives $\omega^2 = k/m$, where ω is measured in radians per second. A circular function repeats itself in 2π radians, so one cycle of vibratory motion is completed when

$$\omega T = 2\pi$$

where T is the time for one complete cycle or T = the period of motion.
The frequency, f, in cycles per second would then equal $1/T$, or $f = \omega/2\pi$.

In addition, acceleration = d^2y/dt^2, and from Newton's second law,

$$F = ma \quad \text{and} \quad (\text{acceleration}) = F/m.$$

Combining these gives-- $F/m + (k/m)y = 0$ or $F = -ky$
where the minus sign is explained by the fact that the force is opposite in sense to the displacement, y, of the mass.
The force acting on the mass in the figure would equal

$$\text{minus } (a/b)^2 \cdot k_{sp} y'$$

Where y' is the displacement (extension or compression) of the spring and k_{sp} is the spring constant of the spring.

$$b/y = a/y' \quad \text{so } y' = (a/b)\cdot y$$

and the force acting on the mass,

$$F = -(a/b)^2 k_{sp} \cdot y$$

But force, F, also equals $-ky$
where k is the equivalent spring constant acting on the mass, so

$k = (a/b)^2 \cdot k_{sp}$, $\omega^2 = k/m$, and $f = \omega/2\pi$ $\quad$ $f = (1/2\pi)\cdot\sqrt{(k/m)} = (1/2\pi)\cdot\sqrt{[(a/b)^2\cdot k_{sp}/m]}$

$(a/b)^2 = (20/8)^2 = 6.25$ $\quad$ $f = (1/2\pi)\cdot\sqrt{(6.25\cdot 75/12)} = 0.995$ cycles per second ■

ROTARY VIBRATION

Angular oscillation considers torsional stiffness of the supporting member and the moment of inertia of the vibrating member, rather than mass and linear spring constant. The undamped natural frequency of torsion of the system shown in Fig. 2-17 is given by the relationship:

$f = (1/2\pi) \times \sqrt{(K/J)}$, where K = Joules/radian = stiffness of supporting shaft and J = polar mass moment of inertia of supported member.

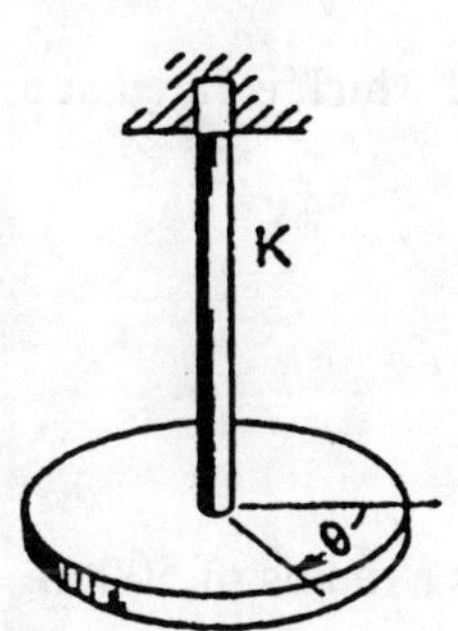

Figure 2-17

EXAMPLE 2-25

A 1.00 kg circular disk 17.0 cm in diameter is rigidly attached to a 9.00 mm diameter steel rod 20.00 cm long. The rod ia fixed at its other end. What is the natural frequency of the system?

Solution:

The torsional spring constant of the rod = GJ_{Rod}/ℓ where:

G = shear modulus, steel = 8.3×10^{10} Pa (FE Handbook)

$J_R = \pi d^4/32 = (\pi \times 0.009^4)/32 = 6.441 \times 10^{-10}$

$K = GJ/\ell = 6.441 \times 8.3/0.200 = 267.3$ Nm/rad

$I_{Disk} = \frac{1}{2}mr^2 = \frac{1}{2} \times 1.00 \times 0.085^2 = 0.00361$ N-m-sec²

$f = (1/2\pi) \times \sqrt{(267.3/0.00361)} = 272.11/6.283 = 43.3$

Frequency = 43.3 cycles per second ■

Problems

2-1. The d'Alembert force is--

a) Force of gravity in France.
b) Force due to inertia.
c) Resisting force due to static friction.
d) Atomic force discovered by d'Alembert.

2-2. The base of a one-kg instrument is set on four rubber mounts, each of which is rated at 5.0 mm deflection per kg of load. What is the natural frequency of vibration?

a) 14 Hz
b) 16 Hz
c) 18 Hz
d) 20 Hz

2-3. A car travelling at 90 km/hr goes around a curve in the road that has a radius of 500 m. What is the lateral force acting on the car?

a) 0.04 g
b) 0.07 g
c) 0.10 g
d) 0.13 g

2-4. A pile driver uses a mass of 500 kg to drive a pile into the ground. If the average resisting force of friction equals 135 kN, the mass of the pile equals 400 kg, and the 500 kg driver drops 6.0 m, which of the following most neatly equals the depth of penetration of the pile? Assume a perfectly inelastic impact between driver and pile.

a) 9 mm
b) 13 mm
c) 17 mm
d) 21 mm

2-5. A horizontal force is applied to a block moving in a horizontal guide. When the block is at a distance x from the origin, the force is given by the equation $F = x^3 - x$. Which if the following most nearly equals the work that is done by moving the block from one meter to the left of the origin to one meter to the right of the origin?

a) 0 J
b) 0.25 J
c) 0.50 J
d) 0.75 J

2-6. In the system shown in Fig. P2-6 the mass, M, equals 10 kg. It is observed to vibrate 10 times in 15 seconds when displaced a small amount from its neutral position. If the mass of the lever arm is disregarded, which of the following most nearly equals the stiffness coefficient of the spring?

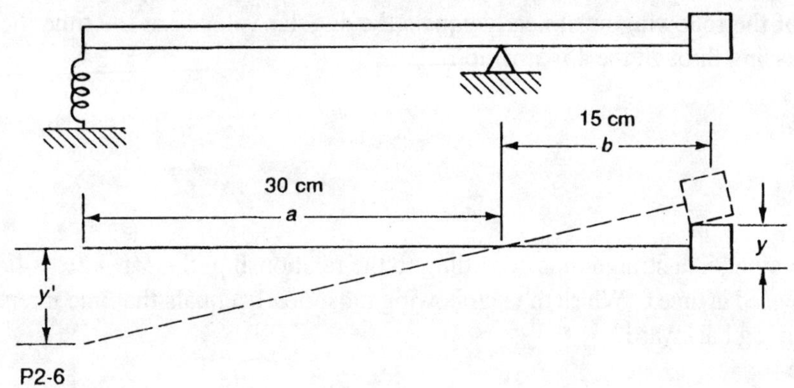

Figure P2-6

a) 57.0 N/m
b) 52.3 N/m
c) 48.2 N/m
d) 43.1 N/m

2-7. An unbalanced force in newtons varies with time in seconds in accordance with the relationship--F = 3,500 - 500 t acts on a body with a mass of 1,500 kg for 5.0 seconds. If the body has an initial velocity in the direction of the applied force of 3.0 m/s, which of the following most nearly equals its final velocity?

a) 9.8 m/s
b) 10.5 m/s
c) 11.2 m/s
d) 11.9 m/s

2-8. A wheel 4.0 m in diameter has a mass of 30.0 kg and a radius of gyration equal to 25.0 cm. If it starts from rest and rolls without slipping down a 30° plane, which of the following most nearly equals its speed when it has rolled 10.0 m measured along the surface of the plane?

a) 11.0 m/s
b) 10.6 m/s
c) 10.2 m/s
d) 9.8 m/s

2-9. A rectangular door one meter wide swings horizontally. It has a mass of 30 kg uniformly distributed. The door is supported by frictionless hinges in stalled along one of its vertical sides. It is controlled by a spring which exerts a torque on it proportional to the angle through which the door is turned from its closed position. If the door is opened 90° and released, the torque due to the spring is then 4.0 J. Which of the following most nearly equals the angular velocity at the time the door has swung back and is passing through the 45° position?

a) 0.61 rad/s
b) 0.69 rad/s
c) 0.79 rad/s
d) 0.85 rad/s

2-10. A body is moving in a straight line according to the relationship $S = \frac{1}{4}t^4 - 2t^3 + 4t^2$ where S equals the distance traveled in time t. Which of the following most nearly equals the time interval during which the body is moving backward?

a) 0 to 2 sec
b) 2 to 4 sec
c) 4 to 6 sec
d) 1 to 3 sec

2-11. A belt 20 cm wide and 6.0 mm thick drives a pulley 30 cm in diameter at a speed of 1650 rev/min. The maximum tension the belt can tolerate is 1.40 MPa. If the slack side has three-fourths the tension of the tight side, which of the following most nearly equals the power that can be transmitted?

a) 7.6 kW
b) 8.7 kW
c) 9.8 kW
d) 10.9 kW

2-12. A kite is 40 meters high with 50 meters of cord out. If the kite is moving horizontally away from the man flying it at a rate of 6.50 km/hr, which of the following most nearly equals the rate at which the cord is being paid out in meters per second?

a) 0.8 m/s
b) 1.1 m/s
c) 1.4 m/s
d) 1.7m/s

2-13. A bullet leaves the muzzle of a gun with a velocity of 900 meters per second and an angle of 45° with the horizontal. Which of the following most nearly equals the maximum distance the bullet will travel horizontally, measured along the same elevation as the gun muzzle? Disregard air resistance.

a) 68 km
b) 73 km
c) 78 km
d) 83 km

2-14. An airplane has a true air speed of 380 km/hr and is heading true north. A 65 km/hr head wind is blowing 45o from true north from the northeast. Which of the following most nearly equals the true ground speed?

a) 337 km/h
b) 351 km/h
c) 366 km/h
d) 380 km/h

2-15. The rifling in an eight mm caliber rifle causes the bullet to turn one rev. for each 250 mm of barrel length. If the muzzle velocity of the bullet is 885 m/sec, which of the following most nearly equals the rate at which the bullet is rotating at the instant at which it leaves the muzzle?

a) 200,000 RPM
b) 204,000 RPM
c) 208,000 RPM
d) 212,000 RPM

2-16. A faster moving body is directly approaching a slower moving body travelling in the same direction. The speed of the faster body is 95 km/hr and the speed of the slower body is 55 km/hr. At the instant when the separation of the two bodies is 30 meters, the faster body is given a constant deceleration. Which of the following most nearly equals the deceleration so that the two bodies just touch at the instant of impact with no shock?

a) 0.21 g
b) 0.25 g
c) 0.29 g
d) 0.31 g

2-17. At the beginning of the drive a golf ball has a velocity of 270 km/hr. If the club remains in contact with the ball for 1/25 sec., which of the following most nearly equals the average force on the ball if the ball has a mass of 45 g?

a) 94 N
b) 89 N
c) 84 N
d) 79 N

2-18. An artificial satellite is circling the earth at a radius of 15,000 km. Which of the following most nearly equals the number of revolutions it makes per day? Assume the radius of the earth is 6,437 km, and assume no air resistance to the travel of the satellite. (The elevation above the earth would equal 8,563 km.)

a) 25/day
b) 36/day
c) 48/day
d) 58/day

2-19. A balloon ia ascending vertically at a uniform rate for one minute. A stone falls from it and reaches the ground in 5.0 sec. Which of the following most nearly equals the height from which the stone fell?

a) 114 m
b) 124 m
c) 134 m
d) 144 m

2-20. A body takes twice as long to slide down a plane at 30° to the horizontal as it would if the plane were smooth. Which of the following most nearly equals the coefficient of friction?

a) $\mu = 0.43$
b) $\mu = 0.50$
c) $\mu = 0.25$
d) $\mu = 0.37$

2-21. Which of the following most nearly equals the distance from the center of a phonograph record turning at 78 rpm that a pickle can lie without being thrown off if the coefficient of friction is 0.30?

a) 5 cm
b) 2 cm
c) 3 cm
d) 4 cm

2-22. A cast aluminum drum 60 cm in diameter and 30 cm in length is supported by an axial shaft on smooth bearings. A light cord wrapped around the drum is pulled with a constant force of 90 N until five meters of cord is unwrapped and is then released. The density of aluminum is 2.70 g/cm³ (as given in the Handbook.) Which of the following most nearly equals the speed of rotation when the pull on the cord has ceased?

a) 78 RPM
b) 89 RPM
c) 98 RPM
d) 67 RPM

2-23. A body with a mass of 90 kg starts from rest at the top of a plane which is at an angle of 60° with the horizontal. After sliding 2.5 meters down the plane, the body strikes a spring with a constant of 17.5 kN/m. The coefficient of friction is 0.25. The body remains in contact with the plane throughout. Which of the following most nearly equals the compression of the spring?

a) 35 cm
b) 40 cm
c) 54 cm
d) 47 cm

2-24. The friction surface on a friction disk type of clutch has an outside diameter of 25.0 cm and an inside diameter of 7.0 cm. The coefficient of friction between the two surfaces equals 0.30. Which of the following most nearly equals the force which must be applied to the plate if the clutch is to transmit 75.0 kW at 3,300 rev/min?

a) 8.2 kN
b) 7.5 kN
c) 6.9 kN
d) 6.4 kN

2-25. A motor car has a maximum speed of 110 km/hr on the level, at which instant the motor is transmitting 45 kW to the wheels. If the tractive effort of the wheels remains constant while the resistance to motion varies as the square of the speed, which of the following most nearly equals the angle of slope up which the car can maintain a speed of 80 km/hr if the mass of the car is 1,800 kg?

a) 1.90°
b) 2.25°
c) 2.75°
d) 3.10°

2-26. A ball, A, has caromed off the cushion at the end of a pool table and is moving at a velocity of 8 m/s in the X direction, see Fig. P2-26. The cue ball hits it at an angle of 20o with a velocity of 15 m/s as shown in the figure. Both balls have the same mass. The coefficient of restitution is 0.60. which of the following most nearly equals the velocity of Ball A after the impact?

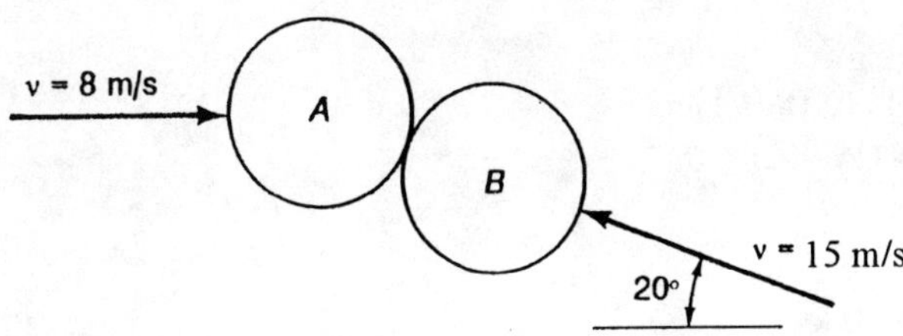

P2-26

Figure P2-26

a) -9.7 m/s
b) -8.5 m/s
c) -7.3 m/s
d) -6.9 m/s

Solutions

2-1. Newton's second law states that Force equals mass times acceleration. Thus when a mass is accelerated there is a resisting, or inertial, force equal to ma. This is termed the d'Alembert force. The correct answer is b).

2-2. The deflection of the four rubber mounts together is:
$0.005/4 = 0.00125$ m
$f = 1/2\pi \times \sqrt{(9.8066/0.00125)} = 14.10$ c.p.s.
The correct answer is a).

2-3. Centrifugal force equals $m \times \omega^2 \times R$
or $F = m \cdot v^2/R = m \times (90/3.600)^2/500 = 1.25 \times m$
or $F = 0.1275$ g
The correct answer is d).

2-4.
$M_1 v_1 = (M_1 + M_2) v_r$ velocity at impact, v_1, equals $\sqrt{(2gh)} = 10.85$ m/s $500 \times 10.85 = 900 \times v_r$
The resulting velocity of the 900 grams equals 6.03 m/sec
The net force acting on the assembly equals 900g - 135 kN = -126.17 kN the rate of acceleration would equal $a = F/m = -126{,}170/900 = -140.2$ m/sec² from $v^2 = v_o^2 + 2as$
we get $0 = 36.36 - 2 \times 140.2 \times s$ which gives $s = 0.130$ m
or K.E. = $\frac{1}{2} Mv^2 = \frac{1}{2} \times 900 \times 36.36 = 16.36$ kJ Work = (net force) × distance or s = K.E./Force
16.36/126.17 = 0.130 meters
The correct answer is b)

2-5. Work = $\int F \cdot ds = \int (x^3 - x)dx$ from ($x = -1$) to ($x = 1$)
Work = $(x^4/4 - x^2/2)$ giving Work = $(\frac{1}{4} - \frac{1}{2}) - (\frac{1}{4} - \frac{1}{2}) = 0$
The correct answer is a).

2-6. Let $a = 30$ cm and $b = 15$ cm and $m = 10$ kg
$d^2y/dt^2 + (k/m)y = 0$ $\omega^2 = k/m$ $F = ma$ See Fig.P2-6
$d^2y/dt^2 = a = F/m$ $F = -ky = -(a/b)k_{sp}y'$ Force on mass = $-(a/b)^2 k_{sp} \cdot y$ $k = (a/b)^2 \cdot k_{sp}$
$\tau = 15$ seconds/10 vibrations = 1.5 s $\tau = 2\pi\sqrt{[(m/k_{sp})(b/a)^2]}$ $k_{sp} = 43.1$ N/meter
The correct answer is d)

2-7. $v = v_o + at$ $a = F/m$ $a = (2.333 - 0.333t)$ m/s²
$\int dv = \int (a\ dt)$ $\Delta v = (2.333 - 0.333t)dt$ from t=0 to t=5
$\Delta v = 2.333t - 0.167t^2 = 7.49$ m/s $v = v_o + \Delta v = 10.49$ m/s
The correct answer is b)

2-8. $I = mk^2 = 30 \times 0.25^2 = 1.875$ kg·m² ΔP.E. = ΔK.E.
ΔK.E. = $\frac{1}{2}mv^2 + \frac{1}{2}I\omega^2 = \frac{1}{2}mv^2 + \frac{1}{2}I(v/r)^2 = v^2(15 + 0.234)$ ΔP.E. = $30 \times 0.5 \times 10 \times g = 1{,}471$ N·m
$v = \sqrt{(1{,}471/15.234)} = 9.83$ m/s
The correct answer is d)

2-9. mass = 30 kg $Tq = I\alpha = k\theta$ where k = spring constant
$I = \int r^2 dm$ from r = 0 to r = 1 $I = \int r^2 dm = \int (r^2 \cdot 30 \cdot dr) = 10.0$ $k = 4.0/1.571 = 2.546$ J/rad
from $\alpha = (d\omega/dt$ and $\omega = (d\theta/t)$ we get $\alpha = \omega(d\omega/d\theta)$ $Tq = -k\theta = -2.546\theta = 10\cdot\omega(d\omega/d\theta)$
$3.928\cdot\int \omega d\omega = -\int \theta d\theta$ $3.928\cdot\omega^2 = -\theta^2 + C$ $\omega = 0$ at $\theta = \pi/2$ So $C = 2.467$
$3.928\cdot\omega^2 = -(\pi/4)^2 + 2.467$ $\omega = 0.686$ rad/sec
The correct answer is b)

2-10. $S = \frac{1}{4}t^4 - 2t^3 + 4t^2$ $v = ds/dt = t^3 - 6t^2 + 8t$
$v = t(t - 4)(t - 2)$ $v = 0$ at $t = 0, t = 4$, and $t = 2$
at t = 3 v = -3 m/s Thus the body moves backward between t = 2 sec and t = 4 sec.
The correct answer is b).

2-11. Max. belt tension = $0.20 \times 0.006 \times 1.40 \times 10_6 = 1{,}680$ N
$Tq = 0.25 \times 1{,}680 \times 0.30/2 = 63.0$ N·m Power = $2\pi\cdot Tq\cdot$rev/sec
Power = $2\pi \times 63.0 \times 1{,}650/60 = 10{,}886$ W or 10.9 kW
The correct answer is d)

2-12. Take L = length of cord and S = horizontal distance along the ground. Then dL/dt = rate at which cord is being paid out. $S_o = \sqrt{(50^2 - 40^2)} = 30$ m 6.50 km/hr = 1.806 m/s
$L = \sqrt{[40^2 + (30 + 1.806t)^2]}$ $d\sqrt{x} = \frac{1}{2}x^{-\frac{1}{2}} dx$
$dL/dt = (108.36 + 6.524t)/[2\times\sqrt{(2{,}500 + 108.36t + 3.262t^2)}]$
at t = 0 $dL/dt = 108.36/(2 \times \sqrt{2{,}500}) = 1.084$ m/s
The correct answer is b)

2-13. Height of apogee, h, equals $v_y^2/2g = (900\cdot\sin 45°)^2/2g$ h = 20,648.5 m
time to reach apogee equals--h/(ave vert velocity) = 20,648.5/[(900·sin 45°)/2] = 64.892 s
horizontal travel = 2 × 64.892 × 900 × cos 45° = 82,594 m
The correct answer is d)

2-14. Backward component of speed 65·sin 45° = 45.96 km/hr
component of speed westerly 65·cos 45° = 45.96 km/hr
Ground speed = $\sqrt{[(380 - 45.96)^2 + 45.96^2]} = 337.2$ km/hr
The correct answer is a)

2-15. Velocity equals 885,000 mm/sec bullet makes one rev in 250 mm
885,000/250 = 3,540 rev/sec or 212,400 rpm
The correct answer is d)

2-16. Use relative velocities. $U^2 = U_o^2 + 2as$ 95 - 55 = 40 km/hr = 11.111 m/sec
$0 = 11.111^2 + 2\cdot a\cdot 30$ $a = -2.058$ m/sec² or 0.210 g
The correct answer is a)

2-17. Impulse = momentum F·t = Mass·v v = 270,000/3,600 = 75.0 m/s
M = 45/1,000 = 0.045 kg F = 0.045·75.0/(1/25) = 84.375 N
The correct answer is c)

2-18. The force outward equals $\omega^2 Rm_s$ This is balanced by the gravitational force inward which equals $k{\cdot}m_s{\cdot}m_e/R^2$ where R equals the distance from the center of mass of the satellite to the center of mass of the earth, k, equals the gravitional constant, m_s is the mass of the satellite, and m_e is the mass of the earth. The value of k can be determined from the acceleration of gravity at the surface of the earth and the radius of the earth, r. $m_o{\cdot}g = k{\cdot}m_o{\cdot}m_e/r^2$ The radius of the earth is approximately 6,437 km or 6,437,000 meters. So k equals $r^2{\cdot}g/m_o = 4.063{\times}10^{14}/m_e$
This gives $\omega^2 = k/R^3 = (4.063/3.375){\times}10^{-7}$ which gives $\omega = 0.003470$ rad/sec
$(0.00347/2\pi) \times 24$hrs $\times(3{,}600$ sec/hr$) = 47.7$ or 48 revolutions about the earth per day.
The correct answer is c)

2-19. Take upward velocity, v, positive. Height, h, of balloon at one minute $h = 60{\cdot}v$ The initial velocity of the stone equals the velocity of the balloon upward. When it is dropped it will be subjected to a negative acceleration of gravity. $60{\cdot}v = 5{\cdot}v - \frac{1}{2}gt^2$ $55{\cdot}v = 122.59$ $v = 2.229$ m/s $h = 60 \times 2.229 = 133.74$ meters (N.B. $g = -9.807$ m/sec^2)
The correct answer is c)

2-20. Force perpendicular to plane = $mg{\cdot}\cos 30° = 0.866{\cdot}mg$
so the frictional force resisting motion equals $\mu{\cdot}0.866{\cdot}mg$. The net force parallel to plane equals $mg{\cdot}\sin 30° - \mu{\cdot}0.866{\cdot}mg = mg(0.500 - 0.866\mu)$ for friction. Net force for no friction = 0.500 mg
$S = v_o t + \frac{1}{2}at^2$ $v_o = 0$ $t_f = 2{\cdot}t_{smooth}$ $\frac{1}{2}a_s t^2 = \frac{1}{2}a_f{\cdot}4t^2$ so $a_f = \frac{1}{4}a_s$ $a = F/m$ and
$F_s = 4F_f$ parallel to surface of plane $0.500{\cdot}mg = 4{\cdot}(0.500 - 0.866{\cdot}\mu)mg$ $\mu = 1.50/3.464$
$\mu = 0.433$
The correct answer is a)

2-21. Force on pickle $=\omega^2{\cdot}Rm$ $\omega = (78/60){\times}2\pi = 8.168$ rad/sec Resisting force $= 0.30{\cdot}mg$
$8.168^2{\cdot}Rm = 0.30{\cdot}m{\cdot}9.807$ $R = 0.0441$ m or 4.41 cm
The correct answer is d)

2-22. Volume $= (\pi/4){\cdot}60^2{\cdot}30 = 84{,}823$ cm^3 Mass = 229.0 kg
$I = \frac{1}{2}MR^2 = 0.5{\cdot}229.0{\cdot}0.30^2 = 10.305$
$KE = \frac{1}{2}I\omega^2 = 90{\cdot}5$ Nm or 450.0 Joules $\omega = \sqrt{(900/10.305)} = 9.345$ rad/sec
$9.345{\cdot}60/2\pi = 89.24$ RPM
The correct answer is b)

2-23. Net force, F, acting down the plane would equal–
$F = mg{\cdot}\sin 60° - mg{\cdot}\cos 60°{\cdot}\ \mu = 654.03$ N
so energy $= 654.03 \times 2.5 = 1{,}635$ J The same can be calculated by calculating the K.E.
$v^2 = 2as$ $F = mg \sin 60° - \mu \times \cos 60° mg = 0.741 \times mg$ or $a = 0.741$ g
$v^2 = 2 \times 2.5 \times 0.741 \times g = 36.334$ K.E. $= \frac{1}{2} mv^2 = \frac{1}{2} \times 90 \times 36.334 = 1{,}635$ J
The total energy to be absorbed by the spring would equal $1{,}635 + s \times 654.03$
where S = distance spring is compressed. $1635 + 654 \times S = \frac{1}{2} \times 17{,}500 \times S^2$
which reduces to—
$S^2 - 0.0747 \times S - 0.1868 = 0$ Solve for S (see Handbook for solution of quadratic equation)
$S = 0.4712$ meter or 47.12 cm
The correct answer is d)

2-24. Watt = joule/sec joule = newton·meter = torque, Tq.
power = $2\pi\cdot$(rev/sec)$\cdot$Tq = watts Tq = Force × radius Force = $\int(\mu p \times 2\pi r\,dr)$
so Tq = $\int[(\mu p \times 2\pi r\,dr)\times r]$
75,000 = Tq·(3,300/60)·2π so Tq = 217.03 N·m for friction clutch
Tq = $\int(2\pi r dr)\mu p r$ where r = radius, μ = friction coefficient, and p = pressure
Tq = $2\pi\mu p\int r^2 dr = 2\pi\mu p r^3/3$ between r = 0.125 and r = 0.035 Tq = 0.001200·p = 217.03 N·m p = 180,858 Pa Clutch face area = $\pi(0.125^2 - 0.035^2) = 0.04524$ m²
Force = 0.04524 × 180,858 = 8,182 N
The correct answer is a)

2-25. Since resistance to motion is proportional to the square of the speed the resistance to motion at 80 km/hr will equal $(80/110)^2$ times resistance to motion at 110 km/hr or 0.5289 × 45 = 23.800 kW This leaves 21.200 kW to overcome the increase in elevation going up the hill. The power required to go up the hill, ignoring the resistance to motion would equal 1,800 × g × sin α × speed
80 km/hr = 80,000/3,600 m/sec = 22.222 m/sec
power requirement = (392,276 × sin α) watts
power available = 21,200 watts so sin α = 21,200/392,276 = 0.0540
α = 3.098°
The correct answer is d)

2-26. Momentum is conserved after an impact. Momentum is a vector quantity, so this means that momentum in the X and Y directions are both conserved. Masses are equal.
Momentum in Y direction = M·15·sin 15° = 3.882M so the velocity of B in the Y direction will remain the same. X direction momentum = $Mu_1 + Mv_1$ = 8 - 15·cos 20° = -6.095 The coefficient of restitution is 0.6 Let: u = A velocity, v = B velocity and initial and final velocities be denoted by subscripts 1 and 2. This gives--$(u_2 - v_{2x})/(-14.095 - 8) = 0.6$ and $u_2 = v_{2x} - 13.257$ But from the conservation of momentum in the X direction--$u_2 + v_{2x} = -6.095$ so $v_{2x} = -u_2 - 6.095$ and $v_{2x} = -v_{2x} + 13.257 - 6.095$ which gives v_{2x} = 3.581 m/s and u_2 = -9.676 m/s
The correct answer is a)

Chapter 3
Thermodynamics

Thermodynamics is defined as the science that deals with the relationship of heat and mechanical energy and the conversion of one into the other. In addition to the mechanical energy--Potential Energy and Kinetic Energy--possessed by solid bodies and dealt with in Mechanics, materials, particularly fluids, and especially gasses, also contain internal energy. It is not possible to determine the absolute amount of energy contained by a substance, neglecting nuclear energy, but the change in energy of a substance is of considerable importance, and that is what is studied in the science of thermodynamics.

The **three laws of thermodynamics** can be described briefly as follows:

The **first law** states that energy can neither be created nor destroyed--this is the law of conservation of energy.

The **second law** states that heat cannot of itself pass from a colder to a hotter system, or, more generally, energy cannot, of itself, pass from a lower to a higher potential.

The **third law** states that the entropy of a substance is zero at absolute zero.

The first two laws form the basis of the study of engineering thermodynamics, and while they cannot be proven true, they are considered valid because all attempts to disprove them have failed.

GENERAL ENERGY EQUATION

From the concept of conservation of energy comes the general energy equation-- $Q - W = \Delta E$ or, heat transferred i.e. heat added to or subtracted from a system, minus work done, either by the system or on the system equals the change in total energy of the system. This can be equated as follows, for a unit mass:

$$Z_1 + u_1 + v_1^2/2 + P_1 \cdot V_1 + Q = Z_2 + u_2 + v_2^2/2 + P_2 \cdot V_2 + W$$

Z_1 and Z_2 are elevations at points 1 and 2 and for a gas the change in potential energy due to the change in elevation is usually negligible, so these terms are generally ignored. The internal energy "u" is often combined with the PV term, pressure times specific volume (Pa × m³/kg) to give enthalpy. $v^2/2$ is the kinetic energy, Q is the heat added, and W is the work done by the fluid.

Take as an example, a closed, non-flow, system a reversible process and a perfect gas. The initial pressure is 500 kPaa (kPa absolute) and it increases to 2,000 kPaa, at a constant temperature, $P \cdot V = C$. The internal energy of the substance increases by 32.00 kJ, and the initial volume is 100 ℓ. Determine the heat added to or subtracted from the substance.

For this case the general energy equation reduces to:

$u_1 + Q = u_2 + W$ since $P_1 \cdot V_1 = P_2 \cdot V_2 = C$ and it is a static, non-flow, system

$P \cdot V =$ constant $V_2 = V_1 \times 500/2{,}000 = 25$ ℓ or 0.025 m³

non-flow work $= \int P\,dV$ $P \cdot V = C$ so $P\,dv = C\,dP/P$

or work done on the substance $= C \times \ln(P_2/P_1)$

$= 500 \times 0.100 \times 1.3863 = 69.315$ kJ

$Q = 32.00 - 69.315 = -37.315$ kJ heat extracted from system.

The FE Handbook lists the basic relationships and it will be available to the examinee during the examination. Thus there is no need to discuss them, it is the application of those relationships that can cause difficulty. So a few examples of the use of thermodynamic principles will be of the most benefit in this review.

FLOW WORK EXAMPLE 3-1.

A fluid at a pressure of 700 kPaa enters a heat engine with a velocity of 190 m/s. At the entrance condition it has a specific volume of 300 ℓ/kg. The fluid leaves the engine with a velocity of 335 m/s and a specific volume of 1.200 m³/kg at a pressure of 35 kPaa. The work output of the heat engine is 2,600 kW, and 85 kW of heat is lost due to radiation and convection. The flow rate of fluid through the engine is 4.50 kg/s. What is the change in internal energy of the fluid?

Solution:

This problem would constitute a number of smaller problems on an FE examination, but it serves to illustrate a number of principles. The problem can be solved with the aid of the general energy equation.

$$Z_1 + u_1 + v_1^2/2 + P_1 \cdot V_1 + Q = Z_2 + u_2 + v_2^2/2 + P_2 \cdot V_2 + W$$

Broken down into parts corresponding to individual exam problems, we have--

The change in potential energy is negligible, so the Z terms can be eliminated.

The change in kinetic energy equals--

$$(v_1^2 - v_2^2)/2 = (190^2 - 335^2)/2 = -38.063 \text{ kJ per kg}$$

The change in PV equals--

$$700 \times 0.300 - 35 \times 1.200 = 168.00 \text{ kJ per kg}$$

Q = 85 kW lost or 85 kJ/s for 4.5 kg/s

$$85/4.5 = 18.89 \text{ kJ per kg}$$

Work done by the heat engine

$$\text{Work} = 2{,}600/4.5 = 577.78 \text{ kJ per kg}$$

Combining the terms gives--

$$u_1 - 38.06 + 168.00 - 18.89 = u_2 + 577.78$$

The reduction in internal energy equals 466.73 kJ/kg

Note: The pressures in this example have been given as absolute values. Ordinarily gage pressures would be given. If gage pressures are given you must add the atmospheric pressure to obtain absolute pressures. Thermodynamic relationships require the use of absolute pressures, and absolute temperatures.

AIR FLOW EXAMPLE 3-2.

A tank with a volume of 2.00 m³ is being filled with air. At one instant of time the temperature of the air in the tank equalled 120 C and the pressure was 1.20 MPa. At this particular instant the pressure was seen to be increasing at the rate of 175 kPa/s and the temperature of the air in the tank was increasing at the rate of 35 C per second. What was the rate of flow of air into the tank a this exact moment in kg/s? Assume air is a perfect gas.

Solution:

The problem asks for the instantaneous flow of air into the tank. This would be dm/dt, or the rate of change of the mass of air in the tank. This can be calculated with the aid of the gas law-- P·Vol = m·R·T. The volume remains constant, as does the gas constant "R", but the mass, pressure, and temperature are all changing with time.

Differentiate the gas law relationship--

$$VdP = mRdT + RTdm$$

dividing through by dt gives--

$V \cdot dP/dt = mR \cdot dT/dt + RT \cdot dm/dt$ and solving for dm/dt gives--

$dm/dt = (V/RT) \cdot dP/dt - (m/T) \cdot dT/dt$

Determine the mass of air in the tank at time zero--

at time zero: T = 120 + 273 = 393oKelvin

P = 1.2 MPa + 101.3 kPa = 1.301 MPaa

$V = 2.00\ m^3$

R = 8,314 J/(kmol·K)

and R_{air} = 8,314/29 J/K = 286.69 J/K

$m = PV/RT = 1.301 \times 10^6 \times 2.00/(286.69 \times 393) = 23.09$

mass of air in the tank at t = 0 equals 23.09 kg

$dm/dt = \{2.00/(286.69 \cdot 393)\} \times 175{,}000 - (23.09/393) \times 35 = 1.0500$ kg/s

EXAMPLE--AIR COMPRESSION 3-3.

Five hundred liters of air is compressed adiabatically from 7.0 kPa. The work done on the air equals 100 kJ and the final temperature is 200 C. Assume the air is a perfect gas. Determine (a) mass of air, (b) the original temperature, (c) the final pressure, (d) the final volume, and (e) the change in internal energy of the air.

Solution:

From the FE Handbook for air:

R = 8,314/29 = 286.69 J/K

c_P = 1.00 kJ/(kg·K)

$k = c_P/c_v = 1.40$

For an adiabatic process $P \cdot V^k$ = constant and the non-flow

Work = $(P_2V_2 - P_1V_1)/(1 - k)$ done by the gas. In this case work is done on the gas so the work is negative. The work done is 100 kJ or 100,000 N·m, so W = -100,000 J

$P_1 = 7.0 + 101.3 = 108.3$ kPaa $\qquad V_1 = 0.500\ m^3$

$P_2 \cdot V_2 = (-100{,}000) \times (-0.4) + 0.500 \times 108{,}300 = 94{,}150$ J

$P_2 \cdot V_2 = m \cdot R \cdot T_2$ so $m = 94{,}150/(286.69 \times 473) = 0.6943$ kg where T_2 = 200 C = 473 K Answer to part (a) is 0.6943 kg

The original temperature is calculated using the gas law--

$T_1 = P_1 \cdot V_1/(m \cdot R) = 108{,}300 \times 0.500/(0.6493 \times 286.69)$

T_1 = 272.04 K or -0.96 C Answer to part (b)

The final pressure can be determined from the relationship for an adiabatic process

$P \cdot V^k$ = constant

$T_2/T_1 = (P_2/P_1)^{(k-1)/k} \qquad (473/272.04)^{3.5} = P_2/P_1$

$P_2 = 108.3 \times 6.9311 = 750.63$ kPaa or 649.33 kPa gage

The answer to part (c) is 649.33 kPa

The final volume can also be calculated from the adiabatic relationship

$P \cdot V^k$ = constant

which gives $T_2/T_1 = (V_2/V_1)^{(k-1)}$

or $V_2 = V_1 \times (473/272.04)^{2.5} = 0.500 \times 3.986 = 1.993\ m^3$

The answer to part (d) is $1.993\ m^3$

The change in internal energy can be calculated in different ways. Since this is a non-flow process the increase in internal energy will equal the work done on the system, or 100 kJ. The increase in internal energy will also equal the increase in temperature times the specific heat at constant volume. $\Delta T = 473 - 272 = 201$ K

the specific heat for air at constant volume from the FE Handbook equals 0.718 kJ/kg·K, and the air mass is 0.6943 kg

$\Delta U = 201 \times 0.718 \times 0.6943 = 100.2$ kJ. The answer to part (e) is 100 kJ

It might be appropriate to mention that $c_v = c_P/k$ and since $k = 1.4$ for air then $c_v = 1.00/1.4 = 0.7143$

ENTROPY

Entropy is a very puzzling subject and there is still disagreement about it among scholars today. There have been many attempts to represent it by physical phenomena without much success. For our purposes it can be considered merely as the integral of dQ/T. It is to be noted that such a relationship implies that entropy is a differential quantity, a change in entropy between two points, though the Third Law implies that there is an absolute entropy for a gas, and such absolute entropies have been calculated for many gasses by calculating the difference in entropy for a as at a given state from zero entropy at absolute zero. There is disagreement about the third law and many contend that it should be stated as: "The change in entropy during an isothermal process approaches zero as the temperature approaches absolute zero."

CARNOT CYCLE

The discussion of entropy leads directly to the Carnot Cycle. The Carnot cycle is described in the FE Handbook and consists of two isothermal and two isentropic processes, and forms a rectangle on the T-S diagram. Heat is added at a relatively high constant temperature, T_1. $\Delta Q_{in} = T_1\Delta S$. Then the working substance expands adiabatically at constant entropy, heat is rejected at the new, lower constant temperature, $\Delta Q_{out} = T_2\Delta S$ and then the substance is compressed to its original volume. The work done equals $Q_{in} - Q_{out}$ or $(T_1 - T_2) \times \Delta S$ since no heat is added or subtracted during an adiabatic process. The efficiency of a Carnot process thus equals $(Q_{in} - Q_{out})/Q_{in}$---work done divided by energy added. This equals $(T_1 - T_2)/T_1$. The significance of this is that this is the maximum efficiency which can be obtained by any thermodynamic engine operating between these to temperatures. Other cycles can be as efficient as a Carnot cycle, but no other cycle can be more efficient. As a result the Carnot cycle is used as a basis of comparison for other cycles and engines.

EXAMPLE 3-4.

An engine operating on a Carnot cycle between temperatures of 500 C and 30 C produces 50 kJ of work.

(a) What is the efficiency of the engine?
(b) How much heat is supplied to the engine?
(c) What change in entropy occurs during the heat rejection portion of the cycle?

Solution:

The upper temperature = 500 + 273 = 773 K. The lower temperature = 30 + 273 = 303 K
Efficiency = (773 - 303)/773 = 0.6080 or 60.80%, Part (a)
Heat supplied = 50.00/0.6080 = 82.24 kJ or 19.66 kcal (b)
The heat rejected equals 82.24 - 50.00 = 32.24 kJ at a constant temperature of 303 K. At constant temperature--
$\Delta S = \Delta Q/T$ so $\Delta S = -32.24/303 = -0.1064$ kJ/°K
or -106.4 kJ/°K or -25.56 cal/°K (c)

EXAMPLE-POLYTROPIC PROCESS 3-5.

Five kg of air at 20 C and one atmosphere is compressed to 700 kPa. The final temperature is 165 C.

Find: (a) The initial volume.
(b) The final volume.
(c) The polytropic exponent.

(d) The work done.
(e) The change of internal energy.
(f) The transferred heat.
(g) The change of enthalpy.
(h) The change of entropy.

Solution:

$P_1 = 101.3$ kPa $\quad T_1 = 20 + 273 = 293$ K
$R = 8{,}314/29 = 286.69 \quad P_2 = 700 + 101.3 = 801.3$ kPa
$V_1 = m \cdot R \cdot T_1/P_1 = 5 \times 286.69 \times 293/101{,}300 = 4.146\ m^3$ (a)
$V_2 = 5 \times 286.69 \times 438/801{,}300 = 0.7833\ m^3$ (b)
For polytropic compression $P \cdot V^n =$ constant, $P_1/P_2 = (V_2/V_1)^n$
$\ln(P_1/P_2) = n \cdot \ln(V_2/V_1) \quad n = -2.0680/(-1.6664) = 1.2410$
(c) The polytropic exponent equals 1.2410
The non-flow work equals $(P_2 \cdot V_2 - P_1 \cdot V_1/(1 - n)$
Work = (627,658 - 420,000)/(-0.2410) = -861.2 kJ (d)
$\Delta U = m \cdot c_v \cdot \Delta T = 5 \times 0.718 \times 145 = 520.5$ kJ (e)
c_v from FE Handbook
The transferred heat = $c_n \cdot \Delta T$
$c_n = c_v \cdot (k - n)/(1 - n)$
$c_n = 0.718 \times (-0.6598) = -0.4737$ kJ/kg·K
Transferred heat = (438 - 293) × (-0.4737) = 68.69 kJ/kg
$\Delta Q = 68.69 \times 5 = 343.45$ kJ (f)
$\Delta H = c_P \cdot \Delta T \cdot m = 1.00 \times 140 \times 5 = 700$ kJ (g)
$\Delta S = m \cdot c_n \cdot \ln(T_2/T_1)$
$5 \times (-0.4737) \times \ln(790/530) = 0.9454$ kJ/oK (h)

PARTIAL PRESSURE

The total pressure in a volume made up of a group of gasses consists of the sum of the partial pressures. Each gas acts as if it alone occupied the total volume. Similarly each fraction of gas can be considered as occupying its volume percent of the total volume at the total pressure.

EXAMPLE 3-6.

A mixture of 40% nitrogen and 60% methane by volume is contained in a 500 ℓ tank. The pressure in the tank is 700 kPa and the temperature is 20 C. It is desired to convert the mixture to 60% nitrogen and 40% methane by venting some of the gas and then adding more nitrogen. How much of the mixture should be vented and how much nitrogen should be added to produce the desired mixture?

Pressure = 700 + 101.3 = 801.3 kPaa $\quad m = PV/(RT)$
Temperature = 20 + 273 = 293 K
For nitrogen $R = 8{,}314/28 = 296.93$ J/kg·K
or $R = c_p - c_v = 1.04 - 0.743 = 0.297$ kJ/kg·K
at state point 1: $P_1 = 0.40 \times 801.3 = 320.52$ kPaa
$T = 293$ K and $V = 0.500\ m^3$
mass = 320.52 × 0.500/(296.9 × 293) = 1.842 kg
For methane $R = 8{,}314/16 = 519.63$ J/kg·K
at state point 1: $P_1 = 0.60 \times 801.3 = 480.78$ kPaa
mass = 480.78 × 0.500/(519.63 × 293) = 1.5789 kg
Total mass of gas at state point 1 = 3.4210 kg

At state point 2 for nitrogen: $P_2 = 480.78$ kPaa

mass = 480.78 × 0.500/(296.93 × 293) = 2.7631 kg

At state point 2 for methane $P_2 = 390.52$ kPaa

mass = 1.0526 kg of methane

Total mass of gas at state point 2 = 3.8157 kg

The mass of methane at state point 2 would be

1.0526/1.5789 = 0.6667 times that at state point 1, so 0.3333 of the initial volume would have to be vented. After venting the mass in he tank would equal 0.6667 × 3.4201 = 2.2808 kg, so 3.8157 - 2.2808 = 1.5349 kg of nitrogen would have to be added to the gas in the tank.

VAPOR PRESSURE

Vapor pressure is the pressure exerted by a vapor when it is in equilibrium with its liquid. Vapor pressure is dependent only on the temperature, it is not affected by the total pressure on the system, no by the amount of space above the liquid. The boiling temperature of a liquid is the temperature at which the vapor pressure is equal to the local atmospheric pressure.

EXAMPLE 3-7.

Five hundred liters of a gasoline is stored in a 1,000 liter steel container at 20 C. The container is sealed and left outside in the sun. On a hot day the temperature of the container rises to 104 C. What would be the gage pressure in the container at the higher temperature? Ignore any expansion of the container.

Solution:

Three factors would affect the pressure in the tank--The gasoline would expand, reducing the volume of the air, the temperature of the air would rise, and the vapor pressure of the gasoline would increase. Assume the coefficient of cubical expansion of the liquid is 0.00139/K. The vapor pressure of the liquid is given as--

10 mm Hg at 20 C and 400 mm Hg at 104 C.

Initially the pressure in the tank is one atmosphere, 101.3 kPa. This total pressure is made up of the vapor pressure of the liquid plus the pressure of the air. The initial vapor pressure, at 20 C, is 10 mm Hg = 1.3 kPa and the vapor pressure at 104 C is 400 mm Hg = 53.3 kPa. The initial pressure in the tank is one atmosphere, 101.3 kPa, and it is made up of the sum of the partial pressure of the air plus the fuel vapor pressure. The partial pressure of the air--

$$P = 101.3 - 1.3 = 100.0 \text{ kPa}$$

The final ullage volume--

$$\Delta V = 0.500 \times 0.00139 \times (377 - 293) = 0.0584 \text{ m}^3$$

$$V_2 = 1.000 - (0.500 + 0.0584) = 0.4416 \text{ m}^3$$

$P_2 = P_1 \times V_1/V_2 \times T_2/T_1 = 100.0 \times 0.500/0.442 \times 377/293 = 145.6$ kPaa air pressure

to this would be added the vapor pressure of 53.3 kPa to give a total pressure of 198.9 kPaa or 97.6 kPa gage.

Problems

3-1.

The expansion of gas during a power stroke in a cylinder of a diesel engine is an example of:

a) An isentropic process.
b) Flow work.
c) Non-flow work.
d) Constant enthalpy.

3-2.

A gas is heated at a constant temperature of 550 C until the entropy increases by 3.00×10^6 J/K. How much heat is added?

a) 2.5 GJ
b) 2.7 GJ
c) 2.9 GJ
d) 3.1 GJ

3-3.

Air is approximately 79% N_2 and 21% O_2 by volume. What is the mass percentage?

a) 79.4% Nitrogen
b) 78.3% Nitrogen
c) 76.7% Nitrogen
d) 75.4% Nitrogen

Solutions

3-1.

Neither the heat nor the entropy remains constant, and there is no flow during the expansion inside a cylinder. It is an example of non-flow work.
The correct answer is c).

3-2.

$\Delta S = \int(dQ/T)$ for a constant temperature process $\Delta Q = \Delta s \times T$
$T = 550 + 273 = 823$ K
so $\Delta Q = 823 \times 3.00 \times 10^6 = 2.469 \times 10^9$ J or 590×10^6 cal

The correct answer is a).

3-3.

For a given volume, V, at a given pressure and temperature, assume the nitrogen occupies a volume equal to $0.79 \cdot V$ and the oxygen occupies a volume of $0.21 \cdot V$.
The mass of each gas will equal

$m = P \cdot V / R \cdot T$

for the assumptions made $P \cdot T$ will be the same for both gasses, so the ratio of the masses will equal

$(V/R)_N/(V/R)_O$

R = 8,314/M.Wt so the mass ratio equals $(0.79 \cdot 28)/(0.21 \cdot 32) = 3.2917$ N_2/O_2
mass percent $N_2 = 3.2917/4.2917 = 76.7\%$ of O_2, 23.3%

The same result can be obtained by assuming each gas occupies the volume by itself. Then the partial pressure of nitrogen would equal 79.0% of the total pressure and the partial pressure of the oxygen would equal 21.0% of the total pressure and V/T would be constant. The mass ratio would equal $(P/R)_N/(P/R)_O$ which gives--
$m_N/m_O = (0.79 \times 28)/(0.21 \times 32) = 3.2917$ as before.

The correct answer is c)

Chapter 4
Heat Transfer

There are three modes of heat transfer--conduction, radiation, and convection. Of these, conduction and radiation are true heat transfer processes since they depend upon the simple existence of temperature differences and the characteristics of the two materials involved in the energy exchange. Convection, on the other hand, depends upon mass transport for the transmission of energy from a higher temperature (higher energy level) to a lower one and so requires consideration of more factors than the other two modes.

CONDUCTION

The rate of conductive heat transfer for steady state, unidirectional flow is given by Fourier's Law of Conduction $Q = kA(T_1 - T_2)/L$

Where: Q = rate of heat flow, J/s or watts

k = thermal conductivity, W/(m·oK)

A = area of surface, meter^2

L = thickness through which heat flows, meter

$T_1 - T_2 = \Delta$ temperature, degrees C or K

It might be noted that thermal conductivities in this country are generally given in Btu/hr·ft·oF it is necessary to multiply the U.S. conventional units by 1.731 to obtain the SI units of W/m·K

EXAMPLE 4-1

What would be the rate of heat flow through a wall constructed of brick and mortar which is 25 cm thick if the wall is 3.0 m by 2.0 m in area and the temperature on one side is 165 C and 55 C on the other side. The coefficient of thermal conductivity is equal to 0.692 W/m·K.

Solution:

$\Delta = 110$ C or K

L = 0.25 m

A = 6.0 m^2

$Q = 0.692 \times 6.0 \times 110/0.25 = 1{,}827$ W or 1.83 kW

If there are a number of thicknesses in series, like a stack of pancakes, the same amount of heat will flow through each thickness and the sum of the temperature drops through the different layers will equal the total temperature differential across the assembly. The temperature drop through one layer, for steady heat flow, would equal--

$\Delta t = Q \times (L/kA)$ $\Delta T = \Sigma\Delta t$ The areas will be equal so $Q/A = \Sigma(L/k)$ or $Q = U \cdot A \cdot \Delta T$ and U = overall heat transfer coefficient $1/U = \Sigma(L/k)$

EXAMPLE 4-2

The wall of a furnace is made up of 22 cm of fire brick, 11 cm of a high-temperature insulating material, next 10 cm of ordinary brick, and 6.5 mm of an asbestos cement board. The inside of the furnace wall is at 980 C, and the temperature of the outside wall is 90 C. What is the rate of heat loss?

Solution:

The thermal conductivities of the different wall components are as follows:

Fire brick k = 1.419 W/m·K

High-temperature insulator k = 0.216 W/m·K

Ordinary brick $k = 0.900$ W/m·K
Asbestos cement board $k = 0.389$ W/m·K

$Q = \Delta T \times U$ Watts/m²

$1/U = 0.22/1.419 + 0.11/0.216 + 0.10/0.900 + 0.065/0.398$

$1/U = 0.791$ $U = 1.264$ $Q = 890 \times 1.264 = 1{,}125$ W/m²

Loss of heat = 1.125 kW/m²

FILM COEFFFICIENT

There is another resistance to the flow of heat through a solid plate or wall from a fluid, often air, to another fluid, and that is termed the film factor or film coefficient. A thin film of practically stationary fluid adheres to the surface of the solid material. The passage of heat through this film is actually a convection process, but for a static condition it can be treated in the same way as conduction through a solid--

$Q = h \cdot A \cdot \Delta T$ where h = the film factor or surface coefficient. The total resistance to heat flow per unit

area per degree then becomes--

$$1/U = \Sigma(1/h) + \Sigma(L/k)$$

EXAMPLE 4-3

If the temperature of the air inside a room is 25 C and the outside temperature is -1.0 C, how much heat would be lost through a window that measured 1.5 m by 2.0 m that was 5.0 mm thick? The thermal conductivity of the glass is 1.731 W/m·K, the film factors on the inside and outside surfaces of the glass pane are 60.75 W/m²·K and 181.70 W/m²·K.

Solution:

The heat transfer coefficient for the glass window is determined by the equation--

$1/U = 1/60.75 + 1/181.70 + 0.005/1.731 = 0.0249$

$U = 40.24$ W/m²·K and $Q = 26 \times 40.24 = 1{,}046$ W/m²

The heat loss through 3.0 m² would equal 3.14 kW

What would the calculated loss be if the film factors were ignored?

Ignoring the film factors gives-- $1/U = 0.005/1.731 = 0.00289$

giving--

$U = 346.2$ W/m² and $Q = 27.0$ kW a very appreciable difference!

If the film factor on the inside surface of the glass equals 60.75 W/m²·K, what is the temperature of the inside surface of the glass?

The same amount of heat will flow through all three thicknesses. The flow through the inner film will equal--1046 W/m²

and the temperature drop across the film will equal-- $1{,}046/60.75 = 17.21$ C .

The temperature of the inside surface of the glass would equal-- $25 - 17.21 = 7.79$ C

Similarly the outside glass surface temperature would equal-- $-1 + 1{,}046/181.70 = 4.76$ C

CIRCULAR PIPES

The flow of heat through pipe walls is a bit different from the flow of heat through flat surfaces. For thin thicknesses, such as thin-walled tubes, the area of heat-transfer is just the area of the wall, but for thick sections the heat-transfer area increases as the radius increases. Using average area as determined by average diameter can give erroneous values, so the logarithmic mean radius is used when one of the annular sections is thick-walled. The greater the ratio of r_2/r_1, (see figure in the FE Handbook) the greater the error introduced by using the arithmetic average to calculate the area of heat transfer.

The logarithmic mean radius, $r_m = (r_2 - r_1)/(\ln[r_2/r_1])$

EXAMPLE 4-4

A thin-walled copper tube 5.0 cm in diameter (k = 350 W/m·K) has a temperature differential of 1.0 C between condensed steam on its inside and stagnate water on its outside. The wall thickness of the tube is 1.50 mm. Both inside and outside are covered with a thin static film having a film factor of 2,850 W/m²·K. What would be the rate of heat transfer per meter of length?

Solution:

The overall heat-transfer coefficient--

$1/U = 1/2{,}850 + 0.0015/350 + 1/2{,}850 =$

$(351 + 4.3 + 351) \times 10^{-6} = 706 \times 10^{-6}$

$U = 1{,}416\ W/m^2{\cdot}K$

The area of tube per meter of length = $\pi \times 0.0500 \times 1.00 = 0.157\ m^2$

$Q = 1.00 \times 0.157 \times 1{,}416 = 21$ watts

N.B. Note the insignificant effect that the resistance of the wall of the tube on the rate of heat transfer. Ignoring it would produce an error in the calculated rate of heat flow of less than one percent. Since the accuracy of heat transfer calculations is generally much less than this, the tube thickness of thin-walled tubes is often ignored.

EXAMPLE 4-5

A 4.0 cm, thin-walled tube is lagged with a 3.5 cm thickness of insulation. (k = 0.071) If the temperature of the tube is 110 C and the temperature of the outside surface of the insulation is 30 C, how much heat would be lost in a 100 m length of pipe?

Solution:

$r_2 = 2.0 + 3.5 = 5.5$ cm $\qquad r_1 = 2.0$ cm

$\ln(r_2/r_1) = 1.0116 \qquad r_m = 3.46$ cm or 0.0346 m

If the resistance of the tube wall is ignored

$1/U = (r_2 - r_1)/0.071 = 0.035/0.071 = 0.493 \quad U = 2.028$

where 0.035/0.071 equals (insulation thickness/k)

Area of heat transfer = $2\pi{\cdot}r_m{\cdot}L = 21.74\ m^2$

$Q = [2\pi{\cdot}L{\cdot}\Delta T \times (r_2 - r_1)/ \ln(r_2/r_1]/[(r_2 - r_1)/k]$ which reduces to--

$Q = 2\pi kL{\cdot}\Delta T/\ln(r_2/r_1)$ as given in the FE Handbook.

$Q = 2\pi \times 0.071 \times 100 \times 80/1.0116 = 3{,}528$ or 3.53 kW

Calculating the rate of heat transfer the long way gives:

$Q = 2.028 \times 21.74 \times 80 = 3{,}527$ watts as before.

RADIATION

The transfer of heat by radiation takes place from a hotter body to a cooler one and is proportional to the fourth powers of the absolute temperatures of the two surfaces. The radiation emitted by a body--

$Q = \epsilon{\cdot}\sigma{\cdot}A{\cdot}T^4 =$ watts

$\sigma = 5.67 \times 10^{-8}\ W/m^2{\cdot}K^4$

ϵ = emissivity of the radiating body

T = absolute temperature, K = °C + 273

A = area, m²

For the radiant heat transfer between two parallel surfaces:

$Q = i_f{\cdot}\sigma{\cdot}(T_1^4 - T_2^4)\ W/m^2$

i_f = interchange factor = $(1/\epsilon_1 - 1/\epsilon_2) - 1$

where ϵ_1 and ϵ_2 are the coefficients of emissivity of the two surfaces.

The coefficient of emissivity equals the coefficient of absorptivity at the same conditions.

The coefficient of absorptivity will vary with the wave-length of the incident light. Window glass for example has a very low absorptivity for short wave-length light such as sunlight. However, it has a very high absorptivity for long wave-length light. This feature enables it to be used for hothouses. Sunlight passes through the glass with little hindrance and is absorbed by the contents of the greenhouse. The contents remain at a relatively low temperature and so radiate long wave-length light to which the glass is almost opaque.

EXAMPLE 4-6

An un-insulated 10 cm O.D. steel pipe runs through a large room with brick walls. The temperature of the pipe is 250 C and that of the brick walls is 2 C. If the emissivity of oxidized steel at 250 C is 0.78, and that of brick at 25 C is 0.89, what is the loss of heat by radiation per meter of pipe length?

Solution:

For a completely enclosed body which is small compared with the size of the enclosing body, the absorptivity of the enclosing body surface can be neglected. So the rate of heat radiated depends only upon the size and characteristics of the pipe in this case.

$$Q = 5.67 \times 10^{-8} \times 0.78 \times \pi \times 0.10 \times 1.00 \times (523^4 - 298^4)$$

$$Q = 930 \text{ watts.}$$

CONVECTION

Natural convection occurs when circulation is caused by differences in density, e.g. when heated water at the bottom of a vessel moves to the top because its density is less than that of the cooler water above it. Forced convection results when circulation is caused by pumps, fans, etc. Calculation of heat transfer by convection depends upon the motion of fluids, thus it is necessary to apply some of the laws of fluid flow to such analysis.

As previously noted there is always a very thin film of essentially stationary fluid on the surface of a solid through which heat is to be transferred. This film always presents the major resistance to the flow of heat from the fluid to the solid or solid to the fluid. As a result, efforts to increase the rate of heat flow by convection are directed toward reducing the thickness of this film. The thickness, and thus the amount of resistance to the flow of heat, can generally be reduced by increasing the velocity of the fluid, especially by changing the flow from laminar to turbulent. However, once the flow has become turbulent, a further increase in velocity will provide little increase in the rate of heat transfer.

The thickness of a film can be estimated, usually within 5 percent, but this requires the use of a complicated equation which includes such factors as density, viscosity, velocity, specific heat, heat transfer coefficient, and size. As a result, it can be expected that required film factors will be supplied in the FE examination.

EXAMPLE 4-7

A heater uses 3.5 cm I.D. by 3.56 mm wall wrought iron pipe for heating a liquid with steam. What would be the effect of increasing the velocity of the liquid so that the liquid film coefficient increases from 1,420 $W/m^2{\cdot}K$ to 1,700? The film coefficient on the steam side equals 11,400 $W/m^2{\cdot}K$, k for wrought iron equals 60 $W/m^2{\cdot}K$

Solution:

$$1/U = 1/1{,}420 + 0.00356/60 + 1/11{,}400$$

giving $U = 1{,}174\ W/m^2{\cdot}K$

For the larger liquid film coefficient--

$1/U = 1/1{,}700 + 0.00356/60 + 1/11{,}400$

giving U = 1,360 W/m²·K a 15.8% increase

What would be the effect on the rate of heat transfer if the steam velocity were increased instead to give a steam-side film coefficient of 17,000 W/m²·K ?

$1/U = 1/1{,}420 + 0.00356/60 + 1/17{,}000$

giving U = 1,216 W/m²·K a 3.6% increase

If instead, the wrought-iron pipe were replaced by a copper tube with the same I.D. and a wall thickness of 3.00 mm, what would be the effect on the rate of heat transfer?

k for copper = 380 W/m·K

$1/U = 1/1{,}420 + 0.0030/380 + 1/11{,}400$

giving U = 1,250 W/m²·K a 6.5% increase

Increasing the steam-side film coefficient by 50%, or increasing the conductivity of the tube wall by 530% produces less increase in the rate of heat transfer than increasing the liquid film coefficient by 20%.

FOULING FACTOR

When heat exchanger tubes become fouled, e.g. encrusted with scale or coated with rust or crud, the rate of heat transfer can be greatly reduced. The resistance of a scale deposit can be determined with the relation-- $1/h_d = 1/U_d - 1/U_c$

where--

h_d = resistance of scale deposit

U_d = overall heat transfer coefficient with scaled tube

U_c = overall heat transfer coefficient with clean tube

EXAMPLE 4-8

What would be the overall heat transfer coefficient for a 2.5 cm I.D. brass tube with a wall thickness equal to 1.25 mm if it were used to heat water using steam on the inside. The film factor on the steam side equals 11,400 and on the water side is 10,200. The scale deposit factors, fouling factors, are 11,400 on each side. For brass k = 109

The mean tube diameter = 0.02625 m tube O.D. = 0.0275 m

Solution:

$1/U = 1/11{,}400 + 1/11{,}400 + 0.00125/109 + 1/11{,}400 + 1/10{,}280$

$U = 2{,}679$ W/m²·K

A more accurate answer could be obtained by calculating the individual temperature drops through the various films--

$1/U = 1/11{,}400 + 1/11{,}400 + 0.00125/[109\cdot(0.02625/0.0275)] + 1/[11{,}400\cdot(0.025/0.0275)] + 1/[10.200\cdot(0.025/0.0275)]$

$1/U = (0.0877 + 0.0877 + 0.0120 + 0.0965 + 0.1078) \times 10^{-3}$

U = 2,553 W/m²·K which is 5% less than the previous calculation.

How much would the rate of heat transfer be improved if the tubes were cleaned on inside and out?

$1/U = (0.0120 + 0.0965 + 0.1078) \times 10^{-3}$

U = 4,623 W/m²·K an 81% increase

If the tubes were cleaned only on the outside--

U = 3,289 W/m²·K a 29% increase

Note that $1/U_d - 1/U_c = 0.0001754 = 1/5{,}701 = 2 \times 1/11{,}400$ where 11,400 are the fouling factors assumed an the inside and outside of the brass heat-transfer tube.

HEAT EXCHANGERS

Any device used to transfer heat can be termed a heat exchanger, but the term "heat exchanger" is usually taken to mean a device wherein one fluid heats, or cools, another. A counter-flow heat exchanger is one in which the hotter fluid enters at one end and the cooler fluid enters at the other end. A parallel flow heat exchanger is one in which both hot and cold fluid enter the same end and the hotter fluid cools as the colder fluid heats. In both cases the temperature differential between the two fluids changes as they pass through the heat exchanger, so the rate of heat transfer in this case cannot be calculated by assuming a constant ΔT, and it has been found that using an average temperature difference gives an erroneous answer. Instead it is necessary to use the LMTD, or logarithmic mean temperature difference, where--

$LMTD = (\Delta T_{max} - \Delta T_{min})/\ln(\Delta T_{max}/\Delta T_{min})$

ΔT_{max} = maximum temperature difference between fluids

ΔT_{min} = minimum ΔT between the two fluids

Problems

4-1.

Radiant heat:

a) Heats homes through hot water pipes embedded in the floor.
b) Cannot transfer through a vacuum.
c) Always depends upon the color of the receiving body.
d) Is not truly a function of the color of the receiving body.

4-2.

What would be the resistance of the scale deposit if the overall heat transfer coefficient for a tube reduced from 1,500 $W/m^2{\cdot}K$ to 1,100 $W/m^2{\cdot}K$?

a) 2500 $W/m^2{\cdot}K$
b) 3500 $W/m^2{\cdot}K$
c) 4125 $W/m^2{\cdot}K$
d) 5550 $W/m^2{\cdot}K$

4-3.

Two different types of heat exchangers are being considered--parallel flow and counter flow. The heating fluid is water at a rate of 4.0 kg/min which enters at 60 C and leaves at 40 C. The heated fluid is also water at a rate of 8.0 kg/min. It enters at 10 C and leaves at 20 C. What would be the relative sizes of the two types of heat exchanger?

a) They would be equal in size.
b) The counterflow heat exchanger would be 10% larger.
c) The parallel flow heat exchanger would be 5% smaller.
d) The parallel flow heat exchanger would be 6% larger.

Solutions

4-1.

Radiant heat transfers from a hotter body to a colder one. For a floor to radiate heat to a human the temperature of the floor would have to be greater than a person's body temperature, whew!. Radiant heat is transmitted to the earth from the sun across many miles of vacuum. Color is a physiological phenomenon which occurs over a fairly narrow range of energy wavelengths. Bodies whose temperatures are above or below the frequency of visible light still radiate or absorb heat. Color, per se, is not a factor in radiant heat transmission.
The correct answer is d).

4-2.

$1/h_d = 1/1{,}100 - 1/1{,}500$ $\qquad h_d = 4{,}125\ \mathrm{W/m^2{\cdot}K}$

The correct answer is c).

4-3.

For parallel flow: $\Delta T_{max} = 60 - 10 = 50$ C
$\Delta T_{min} = 40 - 20 = 20$ C
LMTD = (50 - 20)/ln(50/20) = 30/0.916 = 32.7 C

For counter flow: $\Delta T_{max} = 60 - 20 = 40$ C
$\Delta T_{min} = 40 - 10 = 30$ C
LMTD = (40 - 30)/ln(40/30) = 10/0.288 = 34.7 C

Since the required heat-transfer surfaces are inversely proportional to the mean temperature differences, the parallel flow unit would be 1.06 times as large (6% larger).
The correct answer is d)

4-8

Chapter 5
Fluid Mechanics

The two sub-sections of fluid mechanics are hydrostatics, no fluid motion, and hydrodynamics, fluid flowing with a measurable velocity.

HYDROSTATICS

Fluids consist of both liquids and gasses. Generally speaking the pressure exerted by a gas is considered to be the same throughout its volume. That is the case for a perfect gas; however, for a gas of very high density that may not be accurate. And, thankfully, for extremely large volumes such as the universe, the pressure of the gas, i.e. atmospheric pressure, will vary with its height above the earth or altitude. On the other hand, the pressure at any point in a liquid is equal to the pressure at the surface of the liquid plus the pressure exerted by the head of the liquid--the mass of liquid contained in a unit area vertically above that point times the gravitational constant. Two instances of importance for static liquid pressure are--force on a submerged surface and pressure as measured by a manometer.

(a) **Force on a Submerged Area.** Pressure is non-directional and acts perpendicularly to a surface. Take a surface of width ℓ as shown in Fig. 5-1. The pressure acting at the surface of the liquid will be additive only to the extent that it exceeds the pressure on the back side of the area, this is usually the gauge pressure.

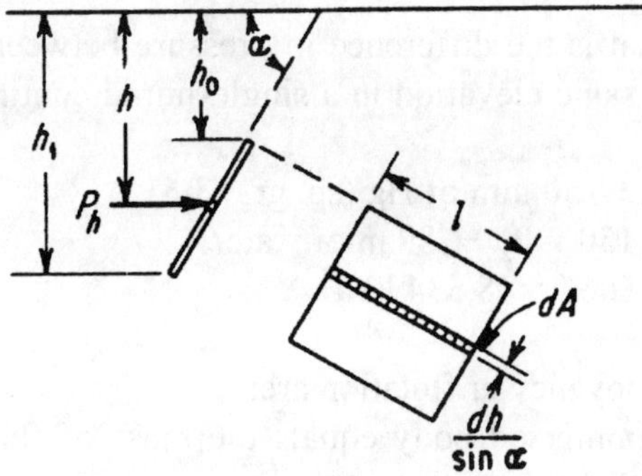

Figure 5-1.

If P = atmospheric pressure the force acting on the submerged area will equal--

$F = [\ell \times (h_1 - h_o)/\sin \alpha \times \rho(h_1 - h_o)/2] \times g = \text{Newtons}.$

Assuming the area is acted on by water, and ℓ = 1.00 m, the top is 1.50 m below the surface, the bottom is 3.00 m below the surface, and $\alpha = 60°$, the force would equal--

$F = 1.7321\ m^2 \times 2.250 \times 1{,}000 \times 9.8066 = 38.219\ kN$

Center of Pressure. The center of pressure of an area is that point about which there would be no moment due to the pressure forces acting on the area. The relationship for calculating the distance "m" from a point "O" to the center of pressure of an area is:

m = I/(A × distance from "O" to centroid of the area)

where A = the area. For this example take point "O" on the surface of the water. The center of pressure would be the same distance below the surface of the water for a vertical plane projection

of the area as for the sloping area.
For the example above, the vertical component of the area would equal $1.00 \times 1.5 = 1.5\ m^2$
$I = bh^3/12 + 1.5 \times 2.25^2$ $I = 7.8751$ so $m = 7.875/(1.5 + 2.25) = 2.333$ so the center of pressure is 0.0830 m vertically below the center of the area or 0.0963 m measured on the surface of the area.

(b) **Manometry**. Manometers are often used to measure the pressure differential between two points.

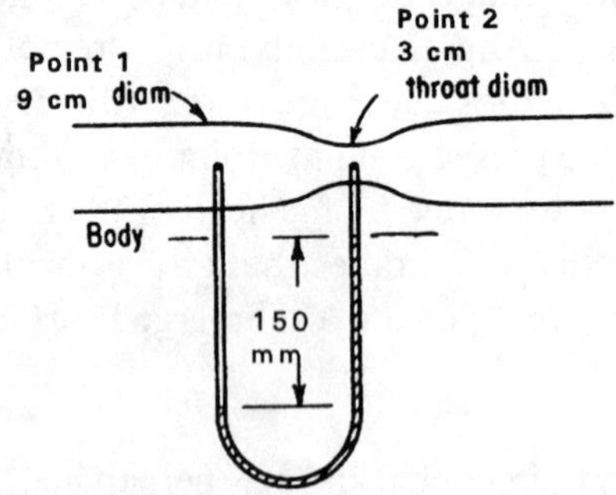

Figure 5-2.

As an example take Fig. 5-2. What is the difference in pressure between points 1 and 2? The pressures at different points at the same elevation in a single-liquid continuous static system are equal. So we have--

$P_1 + 150$ mm of water $= P_2 + 150$ mm of Hg (sp. gr. 13.6)
$P_1 = P_2 + (0.150 \times 13.6) - 0.150 = P_2 + 1.89$ m of water
and $\Delta P = 1.89 \times 1{,}000 \times 9.8066 = 18.534$ kPa

(c) **Buoyancy**. The principles of buoyancy or flotation are:

1. Buoyant force on an immersed body equals the mass of fluid displaced times the gravitational constant, F=ma.
2. A floating body displaces its own mass of fluid.
3. The buoyant force acts up at the C.G. of the displaced volume.

EXAMPLE 5-1

A drum of gasoline, sp. gr. 0.75, is dropped into the ocean, sp. gr. 1.03. The drum is 50 cm in diameter and 76 cm high. Its mass is 125 kg. How much of the drum will extend above the surface if it floats upright?

Solution

The drum will displace--

$125/(1{,}000 \times 1.03) = 0.1214\ m^3$.

The cross-sectional area of the drum = $0.1964\ m^2$, so it will sink 0.6183 m into the ocean. 0.76 - 0.6183 = 0.1417 m or 14.17 cm.

BUOYANCY

EXAMPLE 5-2

A cylinder 1.524 m in diameter and 3.048 m tall is closed at one end. It has a mass of 1,360.5 kg. If the cylinder is placed open end down in water, how far above the surface of the water will the top extend? Assume air and water temperature are constant at 4.444 C.

Solution

General Comments: The gage pressure of the air inside the cylinder times the area of the head must equal the downward force exerted by the cylinder. The pressure of the air will be increased by the reduction of the volume due to the rise of the water inside the cylinder. The floating cylinder will displace an amount of water equal to its mass. If the cylinder were closed it would sink a distance "d". If the bottom were then removed the level of the water in the cylinder would rise, or the open ended cylinder would sink a further distance "d'". The total amount of immersion would then equal d + d'. A distance "h" would thus equal the height of the cylinder minus (d + d').

Area of cap = $\pi/4 \times 1.524^2 = 1.824\ m^2$
Force required to support cylinder = 1360.5 × 9.8066 = 13,342 N
Gage pressure = 13,342/1.824 = 7,315 Pa under cap of cylinder to hold cylinder up.
Total internal pressure = 101,300 + 7,315 = 108,615 Pa
Constant temperature so PV = constant
$V_2 = (101{,}300/108{,}615) \times V_1 = 0.9327\ V_1$
1.0000 - 0.9327 = 0.0673 times height of cylinder
So water will rise 0.0673 × 3.048 = 0.2053 m up into cylinder or the cylinder will sink 0.2053 m into the water.
Pa = N/m^2 water has mass of one g/cc or 1,000 kg/m^3
The cylinder mass of 1,360.5 kg would displace 1.3605 m^3 of water so the cylinder would sink 1.3605/1.824 = 0.7459 m.
If the cylinder were capped on the lower end and placed in water vertically it would sink 0.7459 m. If the cap on the lower end were then removed it would sink a further 0.2953 m to produce an internal pressure of 7,315 Pa gage pressure. The open-ended cylinder would thus sink 0.7459 + 0.2053 = 0.9512 m into the water and the distance h = 3.048 - 0.951 = 2.0968 or 2.10 m

HYDRODYNAMICS

For a closed system with continuous flow of an incompressible fluid (ignoring nuclear reactions) two conditions must be satisfied:

1. The flow at one point must equal the flow at another point or $v_1A_1 = v_2A_2 = Q$--the law of conservation of mass.
2. The energy at one point in the system must equal the energy at another point plus the energy added and minus the energy extracted--the law of conservation of energy.

The first condition is easily met in a closed system. The second condition is satisfied (or determined) by means of the conservation of energy as determined by:

Bernoulli's Equation

$$P_1/\rho g + v_1^2/2g + z_1 + h_A = P_2/\rho g + v_2^2/2g + z_2 + h_L$$

where: P = pressure, Pa
ρ = density, kg/m^3
v = velocity of the fluid, m/s
g = acceleration of gravity, 9.8066 m/s^2
z = elevation, m
h_A = energy added to fluid, m
h_L = energy lost, m

All quantities are in meters of the fluid flowing.

Example 5-3

If the surface of water in a tank is 4.5 m above the bottom of the tank and a 10 mm hole is drilled into the bottom forming a sharp-edged orifice, what would be the initial rate of discharge?

Solution:

Applying Bernoulli's Equation it can Be seen that for state (1), the pressure is atmospheric, the velocity (of the surface of the water) is essentially zero, and no energy is added. For state (2) The pressure is atmospheric and the energy lost will be determined by the discharge coefficient of the orifice. The energy balance then reduces to--

$v_2^2/2g = \Delta z$ or $v = \sqrt{(2 \times 9.8066 \times 4.5)} = 9.395$ m/s

This is the ideal velocity, but a sharp-edged orifice is not 100% efficient and has a discharge coefficient of 0.61, so the actual apparent discharge velocity would be:

$v = 0.61 \times 9.395 = 5.731$ m/s and the rate of discharge would equal 5.731 m/s $\times$ 78.54 mm^2 = 4.501×10^{-4} m^3/s or 27.01 L/min.

Venturi Meter/Orifice Meter

The rate of flow of a liquid is often measured using a Venturi Meter or orifice meter. These are differential-pressure meters and are based on the Bernoulli relationship of constant energy. A Venturi meter is represented in Fig. 5-2. Using the data for the manometer previously calculated-- ΔP = 1.89 m of water, what would the flow rate of water be if the pipe diameter were 9.0 cm, the throat diameter of the Venturi tube were 3.0 cm, and the coefficient of velocity of the flow meter were 0.98? Using Bernoulli's Equation, the elevations at the two points are equal, the difference in the pressure heads is 1.89 m, no energy is added, and the energy lost is determined by the flow coefficient of the meter.

Ideally $(v_2^2 - v_1^2)/2g = 1.89$. But $v_2 = (9/3)^2 \cdot v_1$

so $80\ v_1^2 = 1.89 \times 2 \times 9.8066 = 37.069$

and $v_1 = 0.6807$ m/s ideally or $0.98 \times 0.6807 = 0.6671$ m/s and $Q = 0.66071 \times 0.090^2 \times \pi/4 =$ 0.004244 m^3/s or 4.244 L/s

Calculation of flow rate using an orifice meter is similar, but more complex due to the effect of the location of the pressure taps. The ASME has specifications for flow measurement with orifice meters with different pressure tap locations based on the results of extensive testing and accurate flow measurement with orifice meters depends upon the use of these data.

EXAMPLE 5-4

How many liters per minute (ℓ/min) of gasoline with a sp. gr. of 0.80 are flowing through a 0.0508 m by 0.0254 m throat venturi meter if the differential pressure between the 0.0508 diameter section and the 0.0254 throat diameter section is measured with a mercury manometer, and the height of fluid on the mercury side is 38.1 cm above the height of the gasoline in the other leg?

$c_v = 0.97$ (see figure)

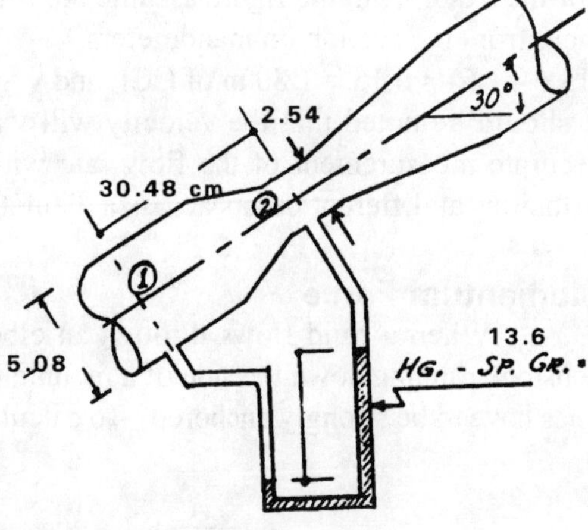

Figure for Example 5-4.

Solution

General comments:

This problem can be solved with the aid of Bernoulli's equation:

$P_1/\gamma_1 + Z_1 + V_1^2/2g = P_2/\gamma_2 + Z_2 + V_2^2/2g$ + head loss due to friction

where the frictional head loss factor is accounted for by the coefficient of discharge, C_v. For this case first ignore the head-loss factor in the equation and the coefficient of discharge and calculate the theoretical velocity. This theoretical velocity must then be multiplied by the velocity coefficient (c_v) to give the actual velocity.

Gasoline density = 0.800 g/cc or 800 kg/m³

mercury density = 13.6 g/cc or 13,600 kg/m³

$P_1 - P_2 = 0.381 \times (13{,}600\text{-}800) = 4{,}876.8$ kgf/m² = 47.825 kPa

$(P_1 - P_2)/\gamma = 4{,}876.8/800 = 6.096$ m of gasoline

$Z_1 - Z_2 = 0.3048 \times \sin 30° = -0.1524$ m of gasoline

$(v_2^2 - v_1^2)/2g = (P_1 - P_2)/\gamma + (Z_1 - Z_2) =$

$6.096 - 0.152 = 5.944$ m

$v_2/v_1 = 2^2/1^2 = 4.0$ so $(v_2^2 - v_1^2)/2g =$

$(16 - 1)v_1^2/(2 \times 9.8066)$

$v_1 = \sqrt{7.7721} = 2.788$ m/sec theoretical velocity

v_1 actual $= 0.97 \times 2.788 = 2.704$ m/sec

$A_1 = 0.0508^2 \times \pi/4 = 2.0266 \times 10^{-3}$

$Q = 2.704 \times 2.0266 \times 10^{-3} \times 60 \times 1{,}000$ l/m³ = 328.80 L/min

Pitot Tube

Pitot tubes are also utilized for flow measurement, such a device is shown in Fig. 5-3. This device operates by measuring the pressure head due to the velocity of the fluid (velocity head).

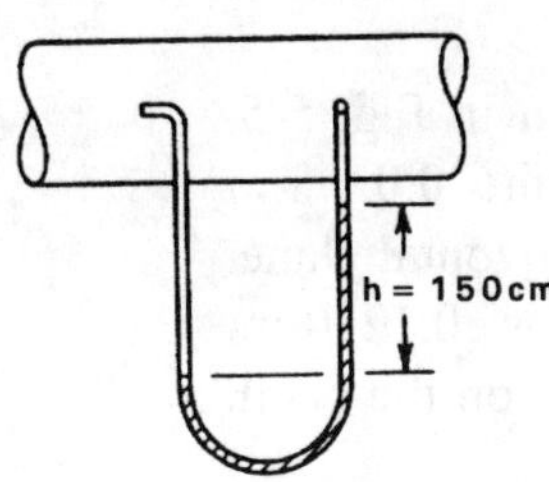

Figure 5-3.

From Bernoulli's equation:

$z_1 = z_2$, $v_2 = 0$, and $(P_2 - P_1)/\rho g = v_1^2/2g$

or $v_1 = \sqrt{(2\,\Delta P/\rho)} = \sqrt{(2gh)}$.

For the example in the figure assume the fluid flowing is CCl_4, sp. gr. of 1.6 then, from the section on manometers:
$(13.6 - 1.6) \times 0.15 = 1.80$ m of CCl_4 and $v = \sqrt{35.304} = 5.94$ m/s velocity at the center of the tube. It should be noted that the velocity will vary across the diameter of the tube, and to obtain an accurate measurement of the flow rate with a pitot tube it would be necessary to measure the velocities at different points across a diameter.

Momentum Force

When a fluid flows through an elbow it exerts a force on that elbow. In the case of a penstock running down the side of a mountain to a turbine, this force can be considerable, and such lines have to be strongly anchored. To calculate the force acting on a bend in a line refer to Fig. 5-4.

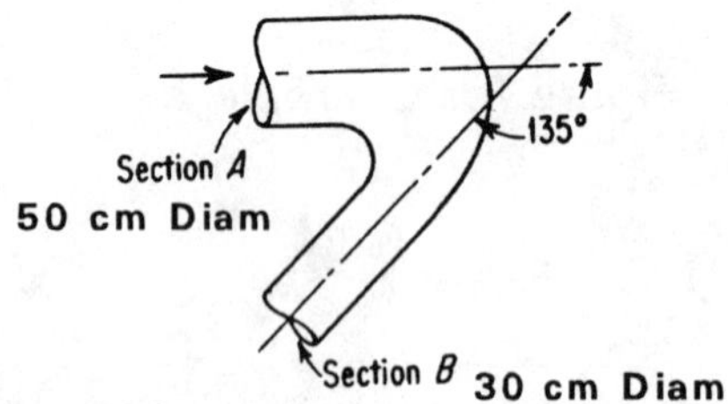

Figure 5-4.

Let the pipe diameter at section A equal 50.0 cm and at section B equal 30 cm. Let the fluid be water and the velocity be 6.00 m/s at section A. The system is horizontal. The flow rate = 0.1964 × 6.00 = 1.178 m^3/s or 1,178 kg/s. The velocity at section B = 6.00 × $(50/30)^2$ = 16.667 m/s.

The momentum forces acting on the elbow are proportional to the changes in velocity since the mass flow is constant in and out of the elbow. In the X direction in the figure
$\Delta v_x = 6 - (-16.667 \times 0.707) = 17.784$ m/s
$\Delta v_y = 0 - 16.667 \times 0.707 = 11.784$ m/s
Force in the X direction = 17.784 × 1,178 = 20.950 kN
Force in the Y direction = 11.784 × 1,178 = 13.882 kN

EXAMPLE 5-5

Water flowing through a 60° bend shown in Fig. 5-5, discharges into the atmosphere through the 0.0508-m diameter exit nozzle. The bend lies in a horizontal plane. The diameter of the entrance section is 0.1270-m. Determine the resultant of the forces acting on the bend. Neglect all losses.

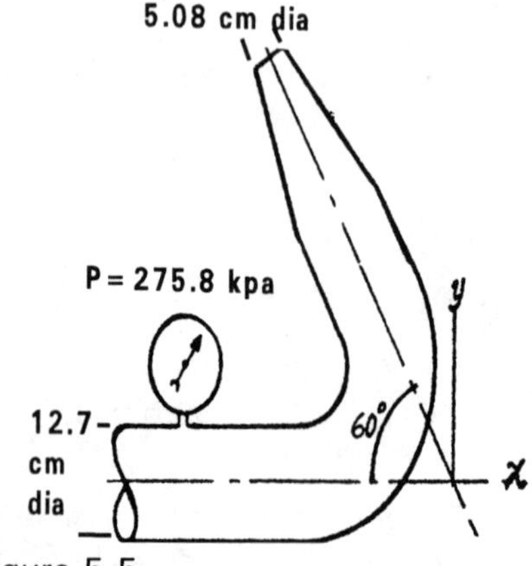

Figure 5-5.

Solution:

General comments:

The fluid will exert a pressure force on the elbow plus momentum forces. Force due to the change in momentum equals mass flow rate times change in velocity. F = (m/dt)Δv . The rate of flow can be calculated using Bernoulli's equation. P_1 = 275.8 kPa gage P_2 = zero gage pressure

$v_2 = (12.7/5.08)^2 \times v_1 = 6.25\ v_1$

$v_2^2 = 39.06\ v_1^2$

$275{,}800/(1{,}000 \times 9.8066) = [(39.06 - 1.00)\ v_1^2]/2g$

$v_1 = 3.807$ m/s

$Q = 3.807 \times 0.01267 = 0.1510\ m^3/s$

0.0508-m diam. gives $A = 2.0268 \times 10^{-3}\ m^2$

0.1270-m diam. gives $A = 0.01267\ m^2$

$275{,}800\ N/m^2 \times (1/1{,}000)\ m^3/kg = 275.8$ Nm/kg

$h = 275.8$ Nm/kg × (1/9.8066)kg/N = 28.124 m of water

$28.124 = (v_2^2 - v_1^2)/2g$ $38.06\ v_1^2 = 551.60$

$v_1 = 3.807$ m/s $Q = 0.0482\ m^3/s$

Pressure force on elbow = 275,800 Pa ×0.01267 = 3,494.4 N

Mass flow rate = 0.0482 m^3/sec × 1,000 kg/m^3 = 48.20 kg/sec

Momentum force on nozzle--

$v_2 = 6.250 \times v_1 = 23.79$ m/s

F = m Δv = 48.20 × 23.79= 1,147 N

$F_y = -1{,}147 \times \sin 60° = -993$ N

$F_x = 1{,}147 \times \cos 60° = 574$ N

Momentum force on elbow mv = 48.20 × 3.807 = 183.5 N

$\Sigma F_x = 574 + 184 + 3494 = 4{,}252$ N

Resultant Force on elbow = $\sqrt{(4{,}252^2 + [-993]^2)} = 4{,}366$ N

Force Exerted by Jet

The force exerted by a jet impinging on a wall (or a rioter) is calculated in a similar fashion. Assume a 2.50 cm diameter nozzle supplied with water at 1.379 MPa. 1,379,000/9,806.6 = 140.62 m of head

$v = \sqrt{(2gh)} = \sqrt{(2 \times 9.8066 \times 140.62)} = 52.517$ m/s

$Q = 52.517 \times 0.025^2 \times \pi/4 = 0.0258\ m^3/sec$ or 25.80 kg/s

F = 52.517 × 25.80 = 1.355 kN if the jet impinges on a wall, v_2 = zero.

Action and reaction are equal and opposite so it would take two or more firemen to hold the nozzle.

Flow Through Conduits

Again referring to Bernoulli's equation it is seen that h_A, the energy to be added to obtain the desired flow rate of fluid is dependent upon, among other things, the magnitude of h_L, the energy lost between the points of interest. This includes energy lost in fittings and frictional losses in the duct. The frictional loss due to flow through a duct is given by the Darcy head loss equation-- $h_f = f(L/D) \times v^2/2g$ where f = friction factor, L = length of the duct, D = diameter of a circular duct, and v and g are as previously defined. For non-circular ducts, D equals four times the hydraulic radius, R_H, where R_H equals the cross-sectional area of flow divided by the wetted perimeter. For a circular pipe $R_H = (\pi D^2/4)/\pi D = D/4$, or four times the hydraulic radius, equals the diameter of the pipe. For a rectangular duct the "D" in the Darcy equation would have to be calculated from the cross-sectional dimensions of the duct. Most calculations of flow of liquids are for circular ducts, or pipes.

The value of "f", the friction factor in the Darcy equation, depends upon the Reynolds Number, and for turbulent flow, the relative roughness of the pipe surface. For Reynolds Numbers below 3,000 flow is laminarand f=64/Re. For Reynolds Numbers above 10,000 flow is turbulent and the friction factor is taken from the Moody Diagram. For Reynolds Numbers between 3,000 and 10,000 the flow is unstable and can be either laminar or turbulent.

The Reynolds Number, Re = $\rho Dv/\mu$ where μ is the absolute viscosity of the fluid usually listed in poises or centipoises, but which must be converted to appropriate units for calculation. Re also equals Dv/ν, where ν = kinematic viscosity measured in stokes or centistokes (also converted to appropriate units). $\nu = \mu/\rho$ v = average velocity in pipe and again $D = 4 \times R_H$ or diameter of a circular pipe.

EXAMPLE 5-6

How much pump power would be required to pump 250 m³/hr of oil of sp. gr. 0.85 with a viscosity of 12 centipoise through 1.0 km of 30 cm I.D. commercial steel pipe? The pipe is horizontal, but it has five 90° elbows with C = 0.85 each.

Solution:

Bernoulli's Equation reduces to $h_A = v_2^2/2g + h_L$ where the head loss includes the frictional loss in the line plus the losses in the elbows.

$A = 0.300^2 \times \pi/4 = 0.07069\ m^2$

$v = (250/3600)/0.07069 = 0.9824\ m/s$

$v^2/2g = 0.9651/19.613 = 0.0492\ m$

So the drop through the five elbows would equal $5 \times 0.85 \times 0.0492 = 0.2091$ m

To determine the friction factor it is necessary to calculate the Reynolds number, $Re = \rho Dv/\mu$.

$\rho = 0.85 \times 1000 = 850\ kg/m^3$ D = 0.300 m v = 0.9824 m/s

The viscosity 12 centipoise = $12 \times 0.001 = 0.012\ Ns/m^2$

N.B. The conversion factor is 1.0 poise = $0.10\ Ns/m^2$

$Re = (850.0 \times 0.300 \times 0.9824)/\ 0.012 = 20{,}876$

From the FE Reference Handbook commercial steel pipe has a roughness of 0.046 mm. The relative roughness for this case thus equals 0.000,046/0.300 = 0.000,153. The Moody Diagram in the Handbook gives f= 0.027 for these values of Re and relative roughness. The frictional head loss due to pipe friction for this case, $f(l/d)(v^2/2g$ would thus equal--

$0.027 \times 1{,}000/0.300 \times 0.0492 = 4.428$ m of oil

Add the five elbows and the velocity head to get a total of $6 \times 0.0492 + 4.428 = 4.723$ m of oil to be supplied to the oil.

$[(250/3600) \times 850]$ kg/s $\times 9.8066 \times 4.732$ m = 2,739 watts or 2.739 kW

EXAMPLE 5-7

Oil (sp. gr. = 0.85 μ = 19 cP) is pumped through 60.96 m of smooth 50.80 mm I.D. hose at 378.5L/min. If the pump has an efficiency of 80 percent, what size motor is required to drive the pump?

Solution:

General comments: The energy which would have to be added to the oil would equal the difference in elevation, plus a velocity head, plus the head loss in the line. (From Bernoulli's equation.)

$P_1/\gamma_1 + Z_1 + V_1^2/2g = P_2/\gamma_2 + Z_2 + V_2^2/2g$ + friction head loss $\Delta Z = 12.19$ m $P_1 = P_2$ = zero gage pressure

$Q = 378.5\ L/min = 0.3785/60 = 6.308 \times 10^{-3}\ m^3/sec$

Area of hose I.D. = $\pi D^2/4 = 2.027 \times 10^{-3}\ m^2$

$v = 6.308 \times 10^{-3}/(2.027 \times 10^{-3}) = 3.112$ m/sec
$v^2/2g = 9.685/(2 \times 9.8066) = 0.494$ m
$Re = \rho Dv/\mu$ $\rho = 1{,}000$ kg/m^3 $\times$ 0.85 sp gr = 850 kg/m^3
$\mu = 19 \times 0.001$ Pa·s (from the FE Reference Handbook)
$Re = 850 \times 0.0508 \times 3.122/0.019 = 7{,}095$
From the Moody Diagram for smooth I.D. pipe f = 0.034 for this Reynolds Number (from the FE Reference Handbook)
$h_L = 0.034\ (60.96/0.0508) \times 0.494 = 20.15$ m
Total head supplied = 12.19 + 20.15 + 0.494 = 32.83 m to a mass of 0.006308 × 850 kg/sec which gives--

$5.3618 \times 32.83 = 176.03$ kg-m/sec

One kgf corresponds to 9.807 Newtons giving--

$176.03 \times 9.807 = 1726$ joules/sec or 1726 watts

$1726/0.80 = 2.16$ kw

PROBLEMS

5-1.

If a pipe is corroded on the I.D. the corrosion will affect the flow through it in the following manner:

a) For a constant pressure supply it will reduce the flow if the flow is laminar.
b) If the Reynolds Number is above 10,000 the corrosion will not have any effect on the flow.
c) For a Reynolds Number of 100,000 it will cause a higher friction factor than for an un-corroded pipe.
d) For a given flow rate the Reynolds number will be higher than for a new, un-corroded pipe.

5-2.

A dam 15 meters high has water behind it to a depth of ten meters. What is the overturning force per meter of width?

a) 1.6 MJ
b) 2.0 MJ
c) 2.3 MJ
d) 1.8 MJ

5-3.

A two-meter I.D. pipe is flowing half full, What is the hydraulic radius?

a) 1.00 m
b) 0.50 m
c) 1.50 m
d) 1.75 m

Solutions:

5-1.

From the Moody Diagram in the FE Handbook it can be seen that the internal roughness of a pipe has no effect on the friction factor for laminar flow. For Reynolds Numbers of 10,000 and above the friction factor is definitely affected by the relative roughness of the inside of the pipe. For a Reynolds Number of 100,000 the Diagram shows that the friction factor would be higher for a rougher inner pipe surface. The internal roughness of the pipe does not enter into the determination of the Reynolds Number.
The correct answer is c).

5-2.

The vertical pressure distribution behind the dam acting on its face will be triangular ranging from zero at the top to--
$P = 1{,}000\ \text{kg/m}^3 \times 10\ \text{m deep} \times 9.8066 = 98.07\ \text{kPa}$ at the base. This will produce a triangle of force acting on the face of the dam equal to--
$F = \frac{1}{2} \times 10 \times 98.07 \times 10^6\ \text{N}$ which will produce a moment of--
$M = 490.35 \times 10/3 = 1.635\ \text{MN/m}$ of width or 1.635 MJ/m
The correct answer is a).

5-3.

The hydraulic radius, R, equals the cross-sectional area of the liquid divided by the wetted perimeter.

Area $= 0.5 \times \pi D^2/4 = 1.5708\ \text{m}^2$
Wetted perimeter $= \pi D/2 = 3.1416\ \text{m}$ $\qquad R = 0.500\ \text{m}$

The correct answer is b)

Chapter 6
Stress Analysis

The subject of stress analysis is usually studied in courses titled "Mechanics of Materials" or "Strength of Materials", and is applied in problems of machine design to determine whether or not the device will withstand the loads to which it may be subjected. Problems of stress analysis include both static and dynamic loading conditions. They may also include environmental conditions such as temperature, corrosion, radiation, and other factors which can influence material properties in the short or long term.

The simplest type of a stress problem is a pair of static forces acting in opposite directions at the longitudinal axis at the ends of a straight bar of constant cross-section. In that case the stress equals the force divided by the cross-sectional area of the bar.

EXAMPLE 6-1 A mass of 100 kg is suspended vertically on the end of a 5.00 mm diameter wire. What is the stress in the wire?
Solution

A one-hundred kg mass exerts a gravitational force of 980.66 N. The cross-sectional area of a 5.00 mm diameter wire equals $19.64 \times 10_{-6}$ m² the stress in the wire equals $(980.66/19.64) \times 10^6$ = 49.93 MPa

STRAIN Stress in a member is always accompanied by strain in the ratio $\epsilon = \sigma/E$ where ϵ = strain, m/m; σ = stress, Pa; and E = Modulus of Elasticity, Pa. In the above example, if the supporting wire is 10.00 m long, how much will it stretch (elongate) when the mass is attached it?

The modulus of elasticity of steel is given in the FE Handbook as 2.1×10^{11} Pa

$\epsilon = 49.93 \times 10^6/2.1 \times 10^{11}) = 237.8 \times 10^{-6}$ m/m

$\delta = 10 \times 237.8 \times 10^{-6} = 237.8 \times 10^{-5}$ m or 2.378 mm

THERMAL STRESS When the temperature of a metal is increased, the metal expands. If a member is constrained from elongating, it will be subject to strain. Since stress and strain are proportional, a strain will result in a stress, or will be evidence of a stress.

The thermal expansion, or compression, of a member is determined by multiplying the thermal coefficient of expansion by the temperature differential. The thermal coefficient of expansion for steel is:

$$\alpha = 11.7 \times 10^{-6} \text{ m/(m}^\circ\text{C)}$$

EXAMPLE 6-2 An exhaust manifold on a diesel engine is held rigidly at both ends. Under extended operation at high power the manifold may become cherry red in color, say 705 C. If the manifold were bolted in place at a temperature of 25 C, what would be the apparent stress in the manifold at the higher temperature? Assume the rest of the engine assembly does not change its dimensions.
Solution

The easiest way to look at a problem like this is to assume that the heated member is free to expand to the higher temperature and then is physically forced to its former dimension. The theoretical longitudinal strain of the steel manifold would equal--

$680 \times 11.7 \times 10^{-6} = 0.00796$ m/m. It would then be compressed an equal amount. This gives an apparent stress of $0.00796 \times 2.1 \times 10^{11}$ = 1.67 gigaPa. This is well above the yield point of structural steel so the manifold would yield in compression.

HOOP STRESS If the wall thickness of a cylindrical vessel is smaller than one-twentieth of the radius it is considered to be a thin-walled vessel. The hoop stress can then be calculated with the equation-- $\sigma = PD/2t$
Where P = internal pressure, D = internal diameter, and t = wall thickness. The load per unit length would equal pressure, N/m^2, times area, D·1.00 , m^2, giving newtons of force. This would be divided by 2·t·1.00, m^2, to give pascals of stress, N/m^2.

Similarly, if the pressure vessel were a closed vessel, the longitudinal force acting on the cylindrical shell would equal the internal area times the pressure, and the longitudinal stress would equal--

$(P \times D^2 \times \pi/4)/(\pi \times D \times t) = PD/4t$ or just half the hoop stress.

If a 750 mm diameter cylinder with a 5.00 mm thick wall is pressurized to 1.5 MPa, what is the hoop stress?

$\sigma = 1.5 \times 10^6 \times 0.750/(2 \times 0.005) = 112.5$ MPa

The longitudinal stress equals 112.5/2 = 56.25 MPa

These constitute the two principal stresses in the wall of the vessel. The maximum shear stress, from the diagram of Mohr's Circle in the FE Handbook, equals one-half the algebraic difference between the two principal stresses.

$\tau_{max} = (112.5 - 56.25)/2 = 28.125$ Mpa

SHRINK FIT A 1.00 m cast-iron wheel has a 1.00 mm thick ring shrunk onto it as a tire. The interference fit is 0.1000 mm. What is the stress in the ring? What is the contact pressure?

The diametral strain is 0.100/1,000 = 0.000100 m/m.

Stress = $E \cdot \epsilon = 2.1 \times 10^{11} \times 10^{-4} = 21$ MPa

Hoop stress $\sigma = PD/2t$ so $P = 2t\sigma/D$

Contact pressure = $2 \times 0.001 \times 21 \times 10^6/1.00 = 42$ kPa

TORSION The angle of twist, ϕ, in a shaft transmitting power equals TL/JG where:

T = torque, N·m
L = length of shaft, m
J = polar moment of inertia = $\pi R^4/2 = \pi D^4/32$, m^4
G = shear modulus of shaft material, Pa
ϕ = radians
degrees = radians $\times 180/\pi$ = radians $\times$ 57.30

EXAMPLE 6-3 A motor running at 1,750 rpm transmits 10 kW of power through a 25.0 mm diameter shaft 2.0 m long. What is the angle of twist in the shaft?

Solution

Watt = J/s = N·m/s = $2\pi NT$ where N is rev/sec
$T = W/2\pi N = 10{,}000/(2 \cdot \pi \cdot 29.167) = 54.57$ N·m
$J = \pi \cdot 0.025^4/32 = 3.835 \times 10^{-8}$ m^4
$G = 8.3 \times 10^{10}$ Pa from the FE Handbook
$\phi = 54.57 \times 2.00/(3.835 \times 8.3 \times 100) = 0.0343$ radians
Angle of twist = 1.9647 degrees

What would the shaft diameter have to be to reduce the angle of twist to 1.00 degree? The polar moment of inertia would have to be increased in the ratio of 1.9647:1.000 so the new shaft diameter would equal $25 \times 1.9647^{0.25} = 25.0 \times 1.1839 = 29.60$ mm

EXAMPLE 6-4 Given the bi-metal shaft shown in Fig. 6-1. If a torque of 2.00 kJ is applied at the point of juncture, and both ends are restrained at their ends, what would be the reaction where the steel section is attached to the wall? What would be the stress in the aluminum section?

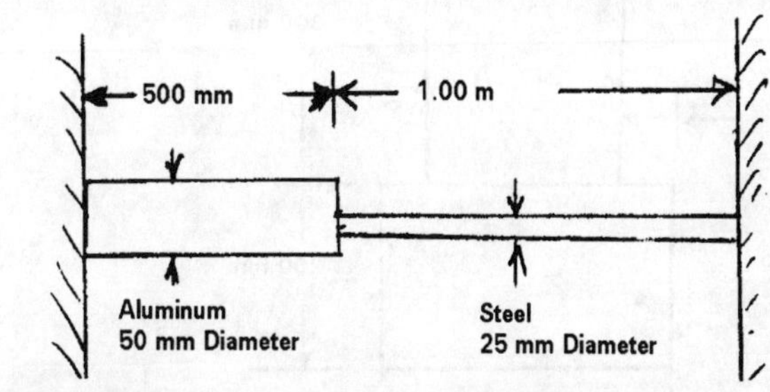

Figure 6-1.

Solution

From the FE Handbook the total angle of twist of a shaft subjected to a torque
$\phi = TL/GJ$
where--

T = torque, joules; L = length of shaft, meters;
G = shear modulus, Pa; and J = polar moment of inertia

The total angle of twist would be the same for both the steel and the aluminum sections.
For the aluminum section:

$J = \pi D^4/32 = \pi \cdot 0.050^4/32 = 6.136 \times 10^{-7}\ m^4$
$L = 0.500$ m
$G = 2.8 \times 10^{10}$ Pa
$\phi = T_{Al} \times 2.910 \times 10^{-5}$

For the steel section:

$J = 3.835 \times 10^{-8}\ m^4$
$L = 1.000$ m
$G = 8.3 \times 10^{10}$ Pa
$\phi = T_{Steel} \times 31.42 \times 10^{-5}$

The total angels of twist are equal so,

$T_{Steel} = 2.91/31.42\ T_{Al} = 0.0926\ T_{Al}$
The sum of the torques in the steel and the aluminum sections equals the total applied torque.
$1.0926\ T_{Al} = 2{,}000$ J $\quad T_{Al} = 1831$ J
$T_{Steel} = 2{,}000 - 1{,}831 = 169$ joules

Stress due to applied torque $\sigma = Tc/J$ For the aluminum section
$\sigma = 1831 \times 0.025/(6.136 \times 10^{-7} = 74.6$ MJ

BEAMS

The stress in a beam, $\sigma = Mc/I$ Where:0

M = moment at the point of interest, usually the maximum moment, N·m or joules
c = distance from neutral axis to outermost fiber
I = moment of inertia

EXAMPLE 6-5 What is the moment of inertia about its centroidal axis of a beam composed of two 200 mm by 300 mm beams connected in the form of a tee, see Fig 6-2.

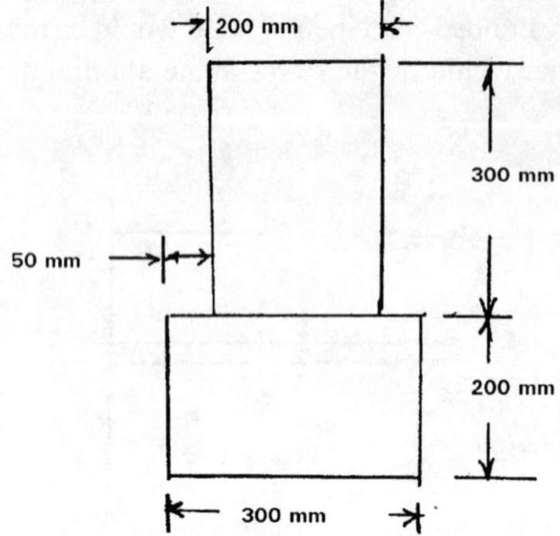

Figure 6-2.

Solution

First it is necessary to determine the location of the centroidal axis. Take moments of the areas about the top of the composite beam and divide by the total cross-sectional area.

(0.150 × 0.060 + 0.400 × 0.060)/(0.060 + 0.060) = 0.275

The centroid of the composite beam is 0.275 m down from the top.

The moment of inertia of the vertical member about its centroidal axis parallel to the axis though the centroid of the composite beam is $I = bh^3/12 = 450\ \mu m^4$ and about the new axis = 0.060 × $(0.275 - 0.150)^2 + 450 = 1{,}387.5\ m^{-6}$

The moment of inertia of the cross member about its centroidal axis equals $0.300 \times 0.200^3/12 = 200\ m^{-6}$ and about the new axis $= (0.125^2 \times 0.060) + 200 = 1{,}137.5\ m^{-6}$

The total moment of inertia of the composite beam about its centroidal axis $I = 2{,}525 \times 10^{-6}\ m^4$

EXAMPLE 6-6 What would be the moment of Inertia of an aluminum box-beam shown in Fig. 6-3? What would be the maximum deflection of such a beam if it were uniformly loaded cantilever beam 3.00 m long, with a distributed loading of 100 kg/m?

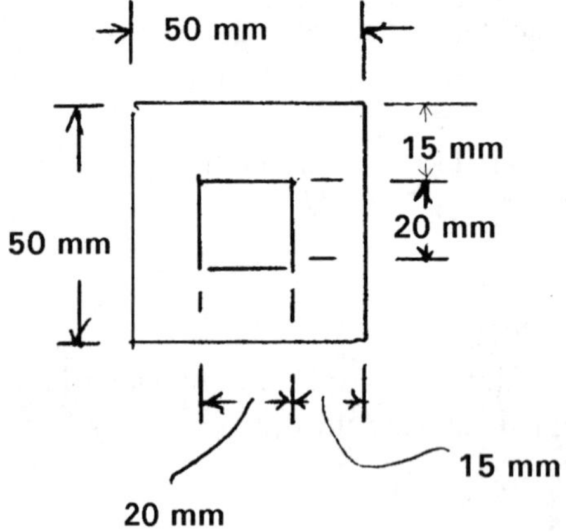

Figure 6-3.

Solution

$I = bh^3/12$ which reduces to $b^4/12$ for a square cross section. For this case the moment of inertia of the section would equal the "I" of the 50 mm square minus the "I" of the 20 mm core section.

$I = (0.050^4 - 0.020^4)/12 = 5.075 \times 10^{-7}\ m^4$

The loading would equal $100 \times 9.8066 = 980.66$ N/m

$\delta = (N/m)\cdot L^4/8EI = 980.66 \times 81/(8 \times 5.075 \times 10^{-7} \times 6.9 \times 10^{10})$

$\delta = 0.284$ m

What would be the maximum stress in the beam?

$\sigma = Mc/I = [(N/m)\cdot L^2/2]\cdot c/I \qquad M = 980.66\cdot L^2/2$

$\sigma = 4{,}413 \times 0.025/(5.075 \times 10^{-7}) = 217.4$ Mpa

EXAMPLE 6-7 What concentrated load at the end of the cantilever beam in the above problem would produce the same deflection?

Solution

From the FE Handbook the deflection of the end of an end-loaded cantilever beam equals-- $\delta = PL^3/(3EI)$ and for a uniformly loaded beam-- $\delta = (N/m)\cdot L^4/8EI$ so

$P = (3/8) \times (N/m) \times L = 0.375 \times 980.66 \times 3.0 = 1{,}103$ N or a load of 113 kg as opposed to a uniformly distributed load of 300 kg.

RADIUS OF GYRATION The moment of inertia is defined as $\int r^2 dm$ or $\int r^2 dA$. If all the mass, or all the area, were concentrated at a distance "r" from the point about which the moment of inertia was to be calculated, then "I" would become just r^2m or r^2A and "r" would equal the radius of gyration. For a circular cross section the radius of gyration equals D/4. $I = \pi D^4/64 = AD^2/16$

since $I = k^2A$ where k = radius of gyration $k = D/4$

COLUMNS A column with a high slenderness ratio, L/r, length divided by the radius of gyration of the column, is prone to buckling, whereas a short column can be treated by normal methods. If the L/r ratio is greater than 30 it is safer to treat the column as a "long column" and use Euler's Formula to determine whether or not the loading is critical.

EXAMPLE 6-8 A 50 mm diameter, 3.00 m long, steel rod is pinned at both ends and is unstressed at 20 C. What is the highest temperature to which the bar may be heated before it will buckle? The thermal coefficient of expansion of steel is 11.7×10^{-6}/oC.

Solution

Euler's formula from the FE handbook is

$$P_{cr} = \pi^2 EI/(kL)^2 = \pi^2 E/[k(L/r)]^2$$

For this case $k = 1 \quad L/r = 3.00/(0.0500/4) = 240.0$

$P_{cr} = 9.870 \times 2.1 \times 10^{11}/240.0^2 = 35.98$ MPa

The compressive stress induced in the steel rod due to an increase in temperature will equal

$\Delta T \times 11.7 \times 10^{-6} \times 2.1 \times 10^{11}$ Pa

so $\Delta T = 35.98/2.457 = 14.64$ C so $T = 34.64$ C

RIVETED JOINTS

Riveted joints can be generally divided into two general classes--joints in which the line of action of the applied force passes through the centroid of the rivet pattern and those in which the force exerts a moment on the rivet pattern. In a simple continuous joint such as is shown in Fig. 6-4A the load is assumed to be distributed equally among the rivets making up the joint. In the sample shown one-quarter of the load would be assumed to be held by each rivet. The small moment force due to the

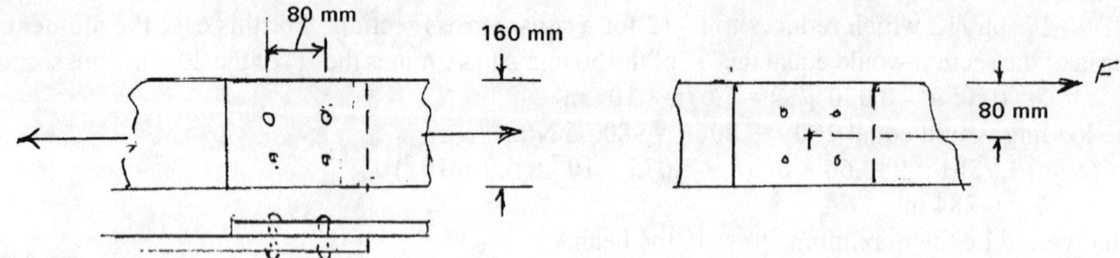

Figure 6-4A   Figure 6-4B

displacement of the lapping plates is ignored. However, if the load is offset from a line of action passing through the centroid of the rivet pattern as shown in Fig. 6-4B, the force due to the effect of the moment must also be included.

EXAMPLE 6-9 The rivets in Fig 6-4B are 12.0 mm in diameter and form a square pattern with sides of 80 mm located centrally in the 160 mm wide plates. The 80 mm is measured between the centers of the rivets at the corners of the square. The load is shown applied at the edge of the 160 mm wide plate. This load will apply a moment load to the rivet pattern equal to $F \times 0.080$ which will be resisted equally by each of the rivets. Each rivet is $0.040 \times \sqrt{2} = 0.05657$ m from the centroid

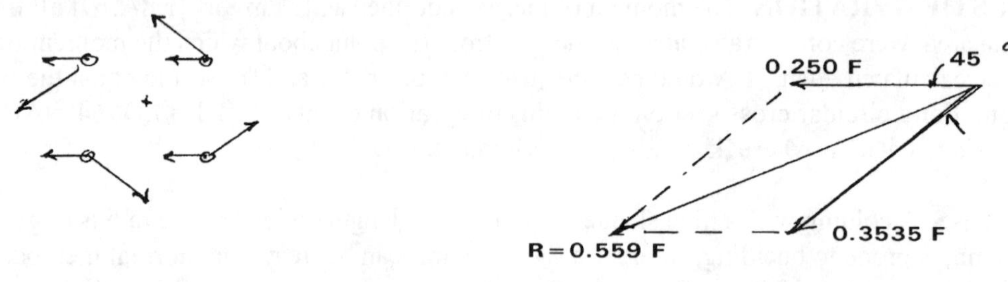

Figure 6-4C.   Figure 6-4D

of the rivet pattern.

The resisting force required to counteract the moment would equal--

$$0.080 \cdot F/(4 \cdot 0.05657) = 0.3535\ F \quad \text{on each rivet. (Fig. 6-4C)}$$

this force would add geometrically to the direct force of F/4 applied to each rivet. This would increase the force on two of the rivets and reduce the force on two others. The maximum force would equal (Fig. 6-4D)

$$[(0.250 + 0.707 \cdot 0.3535)^2 + (0.707 \cdot 03535)^2]^{½} = 0.559\ F$$

The eccentricity of the load application would more than double the maximum stress in the highest stressed rivet in this particular case.

WELDED JOINTS The size of a weld is generally specified by the size of the fillet. For the joint shown in Fig. 6-5 the fillet is given as 6.0 mm. The shear area, however, is the thickness of the throat of the fillet or, for this case: $A = 6 \times 0.707 \times \text{length}$

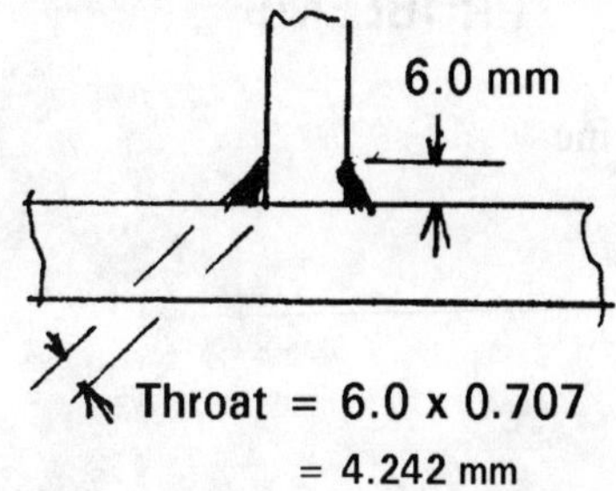

Figure 6-5.

EXAMPLE 6-10 For the joint shown in Fig. 6-5, what is the maximum force which could be applied to the vertical member per centimeter of width if the allowable shear stress of the weld material is 100 MPa?

Solution

For each centimeter length of the joint there would be

$2 \times 0.006 \times 0.707 \times 0.010 = 84.84 \times 10^{-6}$ m² of weld area

$F = 84.84 \times 100.00 = 8{,}480$ N

PROBLEMS

6-1. Mohr's Circle is used to determine:
- a) Shear stress.
- b) Poisson's Ratio.
- c) Residual stress.
- d) Combined stress.

6-2. What would be the maximum stress in a cantilever beam holding a mass of 500 kg at its end if it is 3.00 m long and is made of a 200 mm wide by 300 mm high wooden beam?
- a) 6.8 MPa
- b) 6.1 MPa
- c) 5.5 MPa
- d) 4.9 MPa

6-3. A water tank is filled to a depth of ten meters. It is 35 meters in diameter and is made of steel eight mm thick. What is the stress in the tank wall at the bottom?
- a) 214.5 MPa
- b) 206.1 MPa
- c) 198.8 MPa
- d) 191.9 MPa

Solutions

6-1. Mohr's Circle is used to determine the stress resulting from the stresses on two other surfaces which are perpendicular to each other.
The correct answer is d)

6-2. σ - Mc/I $\quad$ M = 500 × 9.8066 × 3.00 = 14,710 J
$I = bh^3/12 = 0.200 \times 0.300^3/12 = 4.500 \times 10^{-4}\ m^4$
c = 0.150 m
σ = 4.903 MPa
As noted in the FE Handbook the moment of inertia of an area about an axis which is parallel to, but does not pass through its centroid equals its moment of inertia about its centroidal axis plus its area times the square of the distance between the two axes.
The correct answer is d).

6-3.
Hoop stress equals PD/2t
Pressure = 10.0 × 1,000 × 9.8066 = 98.07 kPa at bottom of tank
Stress = 98,070 × 35.0/0.016 = 214.5 MPa
The correct answer is a).

CHAPTER 7
INSTRUMENTATION and MEASUREMENT

Lincoln D. Jones

While measurement is the art of determining the magnitude (or another characteristic of interest) and and expressing its value in relationship to an appropriate standard, instrumentation involves the art of obtaining this information. Instrumentation and measurement is an extremely broad field; a particular device output and the standard it is measured against requires an in-depth understanding of the particular subject area under consideration. For this short review, a number of definitions and the two areas of analog and digital measurements will be briefly discussed.

ANALOG INSTRUMENTATION and MEASUREMENT:

Measurements may be direct (direct measurement implies a measuring device displays or provides the magnitude of the value being measured) or indirect (indirect measurements are usually obtained by calculation or some kind of processing the information). One author* breaks instrumentation into either using "dumb instruments" (where the instrument measures the variable and it is up to the observer to process the data), or "intelligent instruments" (where, after the measurement of the variable is made, further processing and refining the data is made and presented to an observer or to a computer). These intelligent instruments are more likely to be in the digital area (most digital instrumentation is interfaced with analog transducers).

The specific quantity to be measured is referred to as the measurand. To have faith in the results of the measurement of the measurand, one should be concerned with both the accuracy and the precision (they are different, one may have high precision and low accuracy) and the sensitivity of the measurement.

Accuracy is a measure of the difference between a measured value and its actual value ; that is, the degree of agreement between these two quantities.

Precision is the measure of the dispersion (or "spread") of a set measured values about the mean value of the set; that is, the degree of repeatability**

Sensitivity of the measurement is the smallest detectable change in the output of a measuring device for a change in the measurand.**

The system of units of measurements are extremely important and has caused untold conversion problems in the past.. Fortunately the SI ("Systeme International D'Unites") standard is now used almost universally with a few exceptions primarily in the United States. For some areas of engineering, where the SI units are not used, conversion factors are necessary. On the examination, most of the problems should given in SI units; for the few, if any, that do not, the FE Reference Handbook (pages 1,2) will provide you with the conversion factors needed.

Standards:

The primary standards (sometimes referred to as the "absolute standard") in the United States are

*The reference is to: Barney, INTELLIGENT INSTRUMENTATION, Prentice-Hall International, Ltd (UK), 1st ed., 1985, page 1, or 2nd ed., 1988.

**Private notes (for classroom use) of Professor Robert Barksdale, Electrical Engineering Department, San Jose State University, 1992.

maintained by the National Bureau of Standards", NBS. Most calibration services fulfill the obligation of their standards traceable to NBS. Other standards, if listed in order (with the NBS standards being number 1) are: 2) secondary reference standards, 3) interlaboratory standards, and 4) working standards. High quality instruments are calibrated at the time of manufacture and should be updated on a regular basis. Calibration is a function of the chain linking local measurement standards back to the primary standard.

Errors:

For a given accuracy for a direct reading instrument as specified by the manufacturer, say +/- 1% <u>of full scale</u>, the error should be clearly understood. Assuming the instrument is linear over the full scale range, the deviation, d, is defined as the product of the accuracy times the full scale reading and is shown in figure 7.1a; here, one should note that when operating at values less

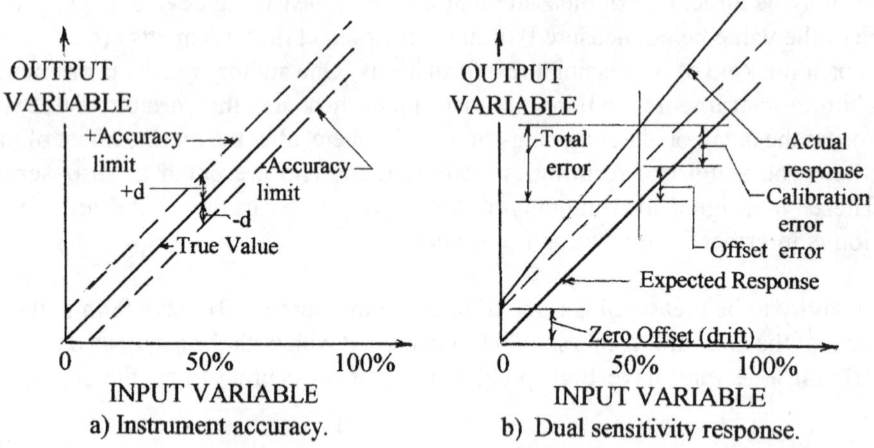

Figure 7.1. Instrument accuracy and effect of dual sensitivity.

than full scale, say 50%, the deviation is the same but the error is doubled (thus, the error would be +/- 2%)! On the other hand, the error can still be greater if there is dual sensitivity. Most instruments are designed to measure one value over a given temperature range, if outside this range, there can be dual sensitivity (the measurand and the temperature or some other factor) and some zeroing offset error as shown in figure 7.1b.

An instrumentation system containing a number of transducers, or a chain of calibrations (say from a secondary standard down to a working standard) introduce linking errors. Chain errors, e_c, corresponds to a given probability that errors produced by a series of links are less than the worst case of each link and adding them together. Statistically, e_c is given as,

$$e_c = \sqrt{e_1^2 + e_2^2 + e_3^2 \ldots e_n^2} \qquad \text{Eq. 7.1}$$

where e_n's are the errors in the different links with each error being the maximum probably for that measurement. A number of influencing factors are present; one of the most important is the direction in which the measurement are made. Some instruments have slightly different readings whether the target reading is approached from above or below; this effect is known as instrument hysteresis.

Analog instruments have a long history as the device used to convert the measurand to numerical values. High accuracy as been attained with direct reading meters and also through "nulling" measurement techniques. This nulling method involves "fine-tuning" or adjusting some measurement standard to have the same value as the measurand and when the difference is zero, the value of the measurand is determined. A typical example is the strain gauge where one balances the resistance of the gauge against a calibrated known resistance of another arm in a bridge circuit. This technique may be automated for fast response time with a mini-control system and an error (or null) detector using instrument amplifiers.

Operational Amplifiers as Instrument Amplifiers:

The operational amplifier (op-amp), where gains may be in the order of 10^6 or greater, the use of the devices, especially for nulling, allow for getting extremely close to the actual null or zero value. Thus the meaning of analog instrumentation frequently requires the understanding of what an op-amp does (although detailed understanding of electronics is not necessarily needed).

For this discussion, the measurement of the measurand will be a conversion of the variable by means of a transducer, usually a voltage, a current, etc., and then processed and then sent on to a display. As an example, one may easily measure and "see" a current wave form by taking the voltage drop across a very low resistance, amplifying (if needed), and taking the output to an oscilloscope. Here an op-amp located at the transducer (the low resistance) may also incorporate noise filters, etc. A typical op-amp configuration that also drives a strip-chart recorder is shown in figure 7.2 with summing capabilities (if all of the switches were on, the output would be the sum of all the signals --

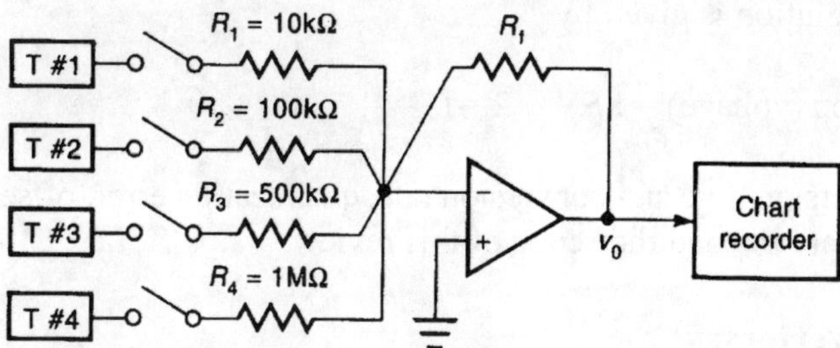

Figure 7.2 An op-amp circuit for matching voltage levels of various transducers. (Caution: only one switch to be closed at a time.)

which is usually not wanted); further assume the sources are several different transducers having different voltage output levels. Note: if one wanted to record the integral of the transducer signal, all that is necessary is to replace the feedback resistor (across the op-amp) with a capacitor as given as,

$$v_o(t) = -\frac{1}{R_1 C}\int_0^t v_1(t)dt + IC, \quad \text{where IC is any initial voltage on C.} \qquad \text{Eq. 7.2}$$

On the other hand, from the output of the transducer, the variable may by be converted to a digital signal, then processed (by the device itself or passed on to a digital computer); this is subject of analog-to-digital conversion that is the topic of the next section.

DIGITAL INSTRUMENTATION and MEASUREMENT:

Most transducers are analog in nature, however, with the advent of inexpensive microprocessors and digital processing along with one-chip analog-to-digital (A/D) and digital-to-analog (D/A or DAC's) converters a great many digital instrumentation systems are in use.

Analog-to-Digital Converters:

an A/D converter, it is. For example if a shaft tachometer (which is usually nothing more than a small dc generator whose voltage output is proportional to speed) is replaced by a binary shaft encoder (see Fig. 7.3), the shaft position is presented as a binary number that may be processed digitally as displacement, speed, acceleration, etc. The output is an n-bit binary (here, 4-bits) code giving a shaft position resolution of $360°/2^n = 360°/2^4 = 22.5°$ (which is typically unacceptable -- most encoders have an n of at least 8-bits or more).For a typical chip A/D converter it has as its input a voltage and for its output a digital code.

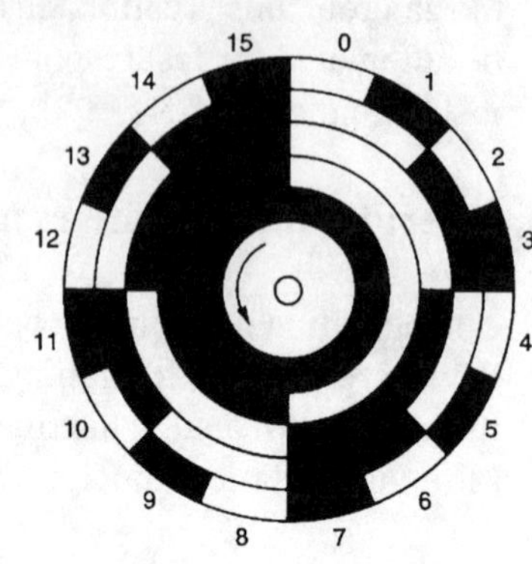

Figure 7.3. A binary shaft encoder.

This device is more complicated than a DAC and is much slower (although a "flash" A/D converter takes only a few hundred ns to operate) than an A/D converter so that it requires a sampler to hold the input voltage long enough to process it (the input appears as a series of steps -- quantization of the input). Here, the resolution is the smallest change in its input analog voltage that can produce a change in a digital code. If the largest input analog voltage to be converted is the full scale voltage, FSV, then the voltage resolution is given by,

$$\text{Resolution (voltage)} = FSV / (2^n - 1) \qquad \text{Eq. 7.3}$$

Other factors such as conversion time and conversion rate, quantization error, offset errors, etc. need to be considered, but they are beyond the scope of this review.

Digital-to-Analog Converters:

The concept of digital-to-analog (D/A or DAC) conversion is relatively simple as compared with A/D conversion. Most D/A converters convert directly from a natural binary code by switching successive weighted data bits to analog values by use of a summing junctions. A typical op-amp circuit is used, although a number of other kind of D/A converters exist, the successive weighted one is easy to understand and will be presented here. Using a summing operational amplifier where each bit is taken to be a weighted (see Fig. 7. 4)bit; for instance if SW(MSB) is ON, then the op-amp output is 1/2 V_{ref}, if only the SW(LSB) is ON for a 4-bit device the output is $1/16V_{ref}$. If V_{ref} is 16 volts, then the output may go (when all SW's are ON) to (1/2+1/4+1/8+1/16)16 = 15 volts. The functional diagram is shown in figure 7.5 without the control lines.

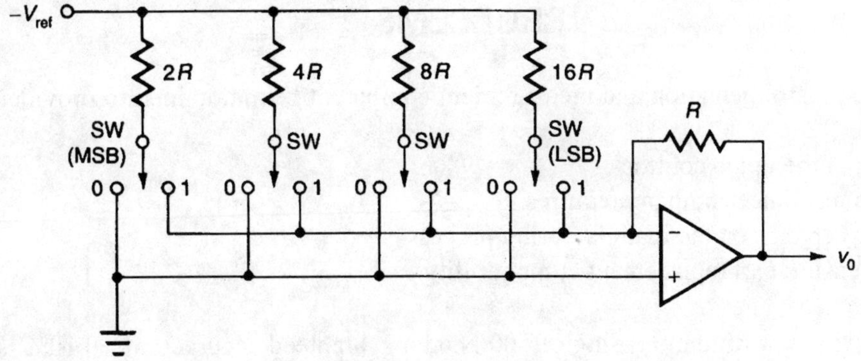

Figure 7.4. Analog portion of a basic D/A converter.

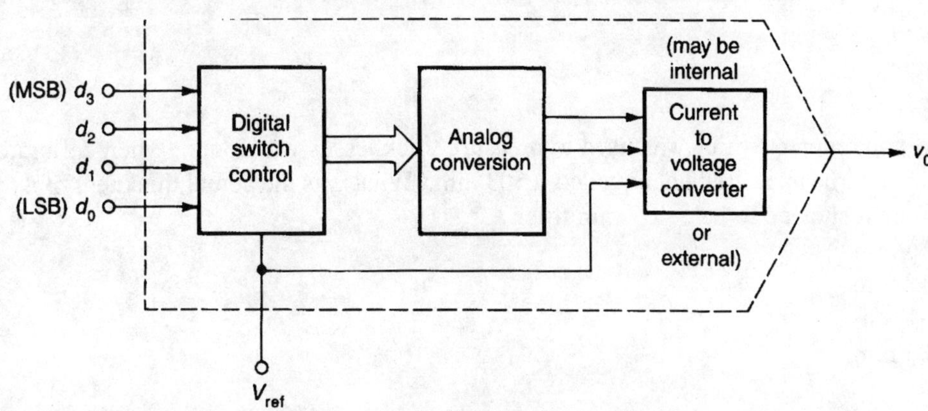

Figure 7.5. Simplified block diagram of a 4-bit D/A converter.

Several of the specifications for a D/A need to be considered:

1) Accuracy - is the difference between the expected value and the actual output value for a given digital input code , this should be less than +/- 1/2 the LSB. As an example, for an 8-bit converter, with a reference voltage of 16 volts, the LSB is 16/256= 0.0625 volts, or the maximum error should be less than 1/2 that or 0.0312 volts.

2) Full Scale - the FS output is the maximum output that may be obtained from the device, that is, if all of the inputs are 1's, then the output is $(2^n-1)/2^n \text{x} V_{ref} = 255/256(16) = 15.94$ volts.

3) Full Scale Range - the FSR is essentially the same as V_{ref}.

4) Resolution - the resolution is the reciprocal of the number of discrete steps (which is a function of the number of bits) and the total number of steps is 2^n-1. Expressed in percent is 0.392% (for the 8-bit converter).

5) Linear Errors - is the deviation of the actual output from a straight line drawn through the idealized output for all steps. A special case is when all of the input lines are zero, is the actual output voltage (the "offset error voltage").

Many instrumentation type problems for an examination may be related to the use of A/D and/or D/A converters. It is suggested that the reader, whether he is proficient or not with digital systems at least have a feel for what these devices do and how they may be interfaced to transducers.

PROBLEMS

7-1. Why must instrumentation and measurement equipment be maintained to provide accurate measurements?

a) It is in the union contract.
b) The insurance company requires it.
c) To keep tabs on the manufacturing process.
d) To ensure part replacement compatibility.

7-2. A scale with a maximum reading of 500 N ad a guaranteed accuracy of ±0.10% is used to weigh amounts of a substance to be sold. A buyer purchases 25.0 kg of a substance. How much could the amount sold be in error?

a) 0.025 kg
b) 0.051 kg
c) 0.062 kg
d) 0.002 kg

7-3. A well worn instrument which was used to measure thicknesses was found, when calibrated, to read zero when the true measurement equalled 0.503 mm. What was the actual thickness of a part which the instrument showed to be 5.409 mm thick?

a) 4.906 mm
b) 5.912 mm
c) 5.661 mm
d) 6.103 mm

SOLUTIONS

7-1. Modern day business is based on the fact that parts must be interchangeable. Before the advent of interchangeability and tolerances, assemblies were made one at a time, and each part was made to fit the particular assembly the artisan was producing. When a part needed to be replaced it had to be made specially. There was no such thing as mass production. Today a business must produce parts to a given tolerance so that a part can be purchased to fit a particular device regardless of where it might be. To accomplish this it is necessary to have a compatible set of dimensions throughout the world.
The correct answer is d)

7-2. 25.0 kg would exert a force of 245.165 N on the scale. The accuracy of ±0.10% applies to the maximum reading of 500 N, or 50.986 kg. The accuracy would thus equal ±0.500 N or 0.051 kg.
The correct answer is b)

7-3. The instrument read zero when the actual measurement was 0.503 mm, so the instrument read 0.503 mm less than the actual dimension. The actual dimension then equalled the instrument reading plus 0.503 mm. The true thickness of the part then equalled--

$$5.409 + 0.503 = 5.912 \text{ mm}$$

The correct answer is b).

Chapter 8
Material Behavior/Processing

Of all the different types of materials used by the mechanical engineer, steel is still probably the most important, especially in the tonnage used.

STEEL

Steel is a heat-treatable alloy, primarily an alloy of iron and carbon. The lower limit for a mild steel is about 0.05 C, though the lower limit might be defined as that value below which iron carbide cannot be thrown out of solid solution, which is below 0.01% . The upper theoretical limit is 1.7% which corresponds to the point in the iron-carbon equilibrium diagram beyond which iron carbide cannot be wholly held in solution at any temperature. The lowest carbon alloys are only slightly, and usually only with difficulty, affected by heat treatment, and are termed "irons" rather than "steels".

Other elements are almost always included in ordinary steel, some by design, and some because it is too difficult or expensive to remove them. Excess amounts of sulphur and phosphorous, for example, are quite deleterious. Too much sulphur results in a condition termed hot-shortness, meaning that the steel becomes brittle at high temperatures. Too much phosphorous, on the other hand, causes cold-shortness.

Steel is not a uniform substance like pure gold, but is made up of components--ferrite and cementite. Ferrite is BCC iron in the crystalline form that exists at room temperature in slowly cooled carbon steel. Cementite is a compound of carbon and iron, Fe_3C. When a carbon steel is cooled at a slow rate from a red heat cementite and ferrite form a laminar composition which resembles mother of pearl and is termed pearlite. Pearlite contains 0.85% C which is the eutectoid composition of steel. A hypoeutectoid steel is one containing less than 0.85% C and when cooled will contain free ferrite. A slowly cooled hypereutectoid steel will contain pearlite and excess cementite.

AUSTENITE

When carbon steel is heated above its transformation temperature the pearlite changes to austenite. Austenite has a face centered cubic lattice and thus contains more atoms per cell, though the packing factor is greater for the FCC cell than for the BCC cell. Austenite is non-magnetic and is more dense than ferrite. Austenite may exist at room temperature only when its transformation as been fully suppressed. Manganese, nickel, and chromium are used to suppress the transformation. Stainless steels, for example, which contain more than 6% nickel and more than 24% of nickel and chromium combined are austenitic at room temperature and are non-magnetic.

MARTENSITE

When austenite is cooled rapidly to a temperature of 200 to 300 C it forms a very hard structure called martensite. This is a skewed tetragonal lattice which forms a needle-like structure. Martensite is less dense than pearlite, so there is a slight increase in volume when a steel is fully hardened. If the cooling is too rapid, the metal does not have time to absorb the increase in volume, and may rupture. Even if the cooling rate is lower than the rate which causes rupture, the increase in volume may still result in severe internal stresses which must be relieved if the part is to be safely loaded. Such parts are heated to a temperature below the lower transformation temperature and "stress relieved".

Elements are added to steel to increase the hardenability, toughness, hardness, corrosion resistance or other property. As an example, plain carbon steel with a BCC structure becomes very brittle at low temperatures. At -70 C, a wrought-iron pipe becomes as brittle as glass and will shatter if stuck by a hammer. But austenitic steels do not exhibit the low-temperature embrittlement phenomenon.

EXAMPLE 8-1

A design for an army tank for use in Antarctica is designed with cleats of high-carbon steel. The ambient temperature is expected to drop as low as -80 C. What should be recommended by a design review?

Solution: Since the tank treads will be in contact with the ground and could reach as low a temperature as the forecast minus 80 C, and will also be subject to continuous shock loading there is a high probability that the treads, being made of a BCC steel would fracture. They should be replaced with an abrasion-resistant non-brittle metal.

NON-FERROUS METALS AND ALLOYS

This classification includes all metals in which iron is not present in large quantities. It includes such metals as copper, aluminum, magnesium, and zinc. Also included are more exotic metals such as silver, gold, tungsten and others which, though used in relatively small amounts, are very important for specialized purposes. Non-ferrous metals are generally used for parts requiring special fabrication and where the ease of fabrication outweighs the higher material costs. Other factors such as density, stiffness, corrosion resistance, electric properties, and color can also affect selection.

MODULUS OF ELASTICITY

The modulus of elasticity is a measure of the elastic deformation of a metal when it is stressed in tension or compression within its elastic limit. It is measured in Pa, and equals pascals divided by m/m elongation.

EXAMPLE 8-2

A 100 kg mass is supported by a 100 m long steel wire 2.5 mm diameter. How much would the wire stretch (strain) elastically?

Solution: The stress in the wire would equal--

$S = 100 \times 9.8066/(4.909 \times 10^{-6}) = 199.8$ MPa

Strain, $\epsilon = 199{,}800{,}000/(2.1 \times 10^{11}) = 95.14 \times 10^{-5}$ m/m

For 100 m length the strain would equal 95 mm

The stiffness of a part can sometimes be very important, and stiffness is a function of the modulus of elasticity, E.

For example the modulus of elasticity for steel, E_s, equals 2.1×10^{11} Pa, while E_a, the modulus of elasticity for aluminum is only 6.9×10^{10}, or 67% less.

EXAMPLE 8-3

A steel cantilever beam holds 150 kg on the end. It is three meters long and ten cm high by five cm wide. To save mass it is proposed to replace the steel beam with an aluminum one of the same length and same proportions. To give the same deflection (a) what would be the deflection of the end of the beam, (b) what would be the dimensions of the aluminum beam, and (c) what would be the percentage mass reduction? The density of steel is 7.9 g/cm³ and of aluminum 2.7 g/cm³.

Solution: The deflection of the beam would equal $PL^3/(3EI)$
For the steel beam $I = b \cdot h^3/12 = 4.167 \times 10^{-6}\ m^4$
$\Delta h = 150 \times 9.8066 \times 3.00^3/(3 \times 2.1 \times 10^{11} \times 4.167 \times 10^{-6})$
$\Delta h = 0.01513$ m or 1.51 cm (a)
For the aluminum beam to have the same stiffness as the steel beam, the product of $E \times I$ for the aluminum beam must be the same as for the steel beam. So $I_{al} = 21/6.9 \times I_s$
$I_{al} = 12.682 \times 10^{-6} = b \times (2b)^3/12$
Which gives $b = (19.023 \times 10^{-6})^{1/4} = 0.06604$
The dimensions of the aluminum for the same stiffness would be 6.604 cm × 13.208 cm (b)
The ratio of the masses would equal--
$6.604 \times 13.208 \times 2.70/(5.0 \times 10.0 \times 7.87) = 0.5985$
There would be a 40 percent reduction in the mass of the beam. (c)

CORROSION

The ability of a metal, or design, to resist corrosion is a very important factor in selecting a particular metal for a specific application. Corrosion causes millions of dollars of structural damage to metal components, and millions more due to surface damage or tarnishing to decorative items. The proper selection of materials or protective systems is very important to a mechanical design.

Corrosion may be defined as the destructive chemical or electrochemical reaction of a material and its environment. The commonest type of corrosion is rust. Rust is a reddish-brown or orange coating on iron or steel due to oxidation of the surface when exposed to air or moisture. It is made up principally of hydrated ferric oxide. Rust is formed by the galvanic action of the iron and small particles of other metals contained in it in contact with acidic solutions formed by the dissolution of carbon dioxide in rainwater and dew. Moisture is always present in the atmosphere, so the potential for electrochemical corrosion always exists.

ELECTROCHEMICAL CORROSION

If two dissimilar metals in contact are joined by an electrolyte, they will form a galvanic corrosion cell and one will be dissolved or corroded. An electrolyte is a substance capable of carrying ions. It is typically an aqueous solution and can be formed from condensed moisture, dew, or rain and contaminated by small amounts of dirt, salts, acids, or alkalis from the atmosphere. The anode and cathode of a corrosion cell can even be parts of the same component with different potentials. The anodic material, the one with the highest potential in reference to hydrogen, as shown in the table of "Standard Oxidation Potentials for Corrosion Reactions" in the FE Handbook, will be corroded. For example if aluminum rivets should be used to join steel sheets, the rivets would be corroded and could fail if not protected. Conversely, if two aluminum plates should be bolted together with iron bolts, pitting due to corrosion would occur in the plates next to the bolts.

It is noted from the table that zinc has a higher oxidation potential than iron, this explains why steel items are sometimes zinc coated--the zinc will corrode in preference to the steel and leave the steel un-corroded. The fact that the cathode is resistant to corrosion is made use of in protecting ship hulls by attaching sacrificial anodes of metals such as magnesium to the hull. Cathodic resistance to corrosion is also made use of in protecting buried pipelines by connecting them to the negative side of a DC current supply.

NON-METALLIC MATERIALS

There are many non-metallic materials of importance to engineers, but most of them lie within the province of civil, chemical, and electrical engineers. However, mechanical engineers will find it of value to be aware of some of the properties of non-metallic materials which can be of benefit in different mechanical designs. Many plastics can prove quite useful to the M.E. in his designs. Plastics have the advantage of being more formable than any metallic material. In addition, plastics can have low density and can have special properties such as specially-desired heat or electrical insulation, corrosion resistance, low friction coefficient, resistance to deterioration by moisture, good color range, or other property not obtainable in the required amount with a metal. A plastic is defined in the broadest sense as any non-metallic material which can be molded to shape. Plastics can be molded, cast, or extruded. They can also be used for coatings and films. Plastics are generally lumped into two general classifications--thermoplastic and thermosetting. Thermoplastics are those which can begin to soften at a temperature as low as 60 C and then can be molded without any change in chemical structure. Thermosetting materials undergo a chemical change when molded and cannot be re-softened by heating. Thermosetting plastics usually require considerably higher mold temperatures than thermoplastics, and the finished part can usually withstand much higher temperatures without deforming.

Problems

8-1. A design engineer should know what a metal's electromotive potential is because it tells him:
- a) If the metal can be heat treated.
- b) What the atomic structure is.
- c) How easy the metal can be formed.
- d) Its resistance to corrosion.

8-2. An extensometer is attached to a 15.0 cm length of a test piece of unknown metal in a laboratory. The test section is one cm square. It was then subjected to a tensile force of 20.0 kN and the elongation of the test section was measured. It was found to have increased in length by 0.008475 mm. What was the modulus of elasticity of the metal?
- a) 21×10^{10} Pa
- b) 6.9×10^{10} Pa
- c) 35.4×10^{10} Pa
- d) $14{,}8 \times 10^{10}$ Pa

8-3. Plastically deforming steel at room temperature--
- a) Increases the ultimate strength.
- b) Increases the corrosion resistance.
- c) Makes it non-magnetic.
- d) Increases the yield strength.

SOLUTIONS

8-1. The electromotive potential of a metal indicates how susceptible it will be to corrosion in an application.
The correct answer is d).

8-2. The modulus of elasticity is measured in pascals, or newtons per square meter. The test showed a strain of--

$\epsilon = 0.008475/150 = 0.000565$ mm/mm for a stress of

$S = 20{,}000/0.0001 = 200$ MPa $\quad S = \epsilon \cdot E$ so--

$E = 20 \times 10^7/(5.65 \times 10^{-4}) = 3.54 \times 10^1$ Pa which is 70% greater than the modulus of elasticity for steel. The unknown metal must be tungsten.

8-3. Cold working, or plastically deforming steel at room temperature increases the yield strength and reduces the ductility.
The correct answer is d).

Chapter 9
Computers and Numerical Methods
Lincoln D. Jones

Introduction

The Fundamentals of Engineering Exam contains seven questions concerning computers in the morning session and three questions in the afternoon session. These questions cover the topics of operating systems, networks, interfaces, spreadsheets, flow charting, and data transmission. Each of the branch specific afternoon exams contain three questions on numerical methods related to that branch.

You should review the glossary of important computer related keywords that is presented at the end of this chapter to ensure you have a basic understanding of this broad general topic. Should you find any term unfamiliar to you, a review of that topic would be warranted. The current exam does not include a programming language such as FORTRAN or BASIC but use of applications such as spreadsheets are included. You should be familiar with one of the popular spreadsheet programs such as Excel, Quattro Pro or 1-2-3.

Number Systems

The number system most familiar to everyone is the decimal system based on the ten symbols 0 through 9. This base 10 system requires ten different digits to create the representation of numbers.

A far simpler system is the binary number system. This base 2 system uses only the characters 0 and 1 to represent any number. A binary representation 110, for example, corresponds to the number

$$1x2^2 + 1x2^1 + 1x2^0 = 4 + 2 + 0 = 6$$

Similarly the binary number 1010 would be

$$1x2^3 + 0x2^2 + 1x2^1 + 0x2^0 = 8 + 0 + 2 + 0 = 10$$

The digital computer is based on the binary system of on/off, yes/no, or 1/0. For this reason it may at times be necessary to convert a decimal number (like 12) to a binary number (for 12 it would be 1100).

Example 1

The binary number 1110 corresponds to what decimal (base 10) number?

Solution

$$1x2^3 + 1x2^2 + 1x2^1 + 0x2^0 = 8 + 4 + 2 = 14$$

Data Storage

Memory chips include random access memory (RAM), read only memory (ROM),and programmable read only memory (PROM).

Diskettes have been the primary way to move data about on personal computers. Initially diskettes were 5-¼ inch "floppies" (360K to 2MB of data) that could be damaged easily. They have since been replaced by 3-½ inch diskettes (720 to several MB of data) in a hard plastic housing.

Hard drives (currently about 500MB to 2GB of data) are units with the hard disks permanently sealed in a module. Other storage devices are removable hard disk cartridges, optical disks, read-only media (ROM), write once, read many media (WORM), and compact disk read-only memory (CD-ROM).

Data Transmission

A computer produces digital signals (represented as on-off or 0-1), but these signals often must be transmitted on voice transmission (analog) systems. The conversion of digital signals to analog signals is called *modulation*. The subsequent conversion back to digital from analog is called *demodulation*. The device doing this conversion is called a *modem* (short for modulation-demodulation). Early modems had speeds of 300-1200 bits per second, but modern modems operate at 14,400-28,800 bits per second.

Data transmission requires that the equipment at both the sending and receiving locations must be able to "talk" to each other. Two methods commonly used are asynchronous and synchronous transmission. In asynchronous transmission a start signal is sent at the beginning and a stop signal at the end of each character. Synchronous transmission, on the other hand, is where a bit pattern is transmitted at the beginning of the message to synchronize the internal clocks of the sending and receiving devices.

Programming

Computer programming may be thought of as a four-step process:

1. Defining the problem
2. Planning the solution
3. Preparing the program
4. Testing and documenting the program

Once the problem has been carefully defined, the basic programming work of planning the computer solution and preparing the detailed program can proceed. In this section the discussion will be limited to two ways to plan a computer program to solve a problem:

Algorithmic Flowcharts
Pseudocode

Algorithmic Flowcharts

An algorithmic flowchart is a pictorial representation of the step-by-step solution of a problem using standard symbols. Some of the commonly used figures are shown in Figure 9-1. Consider the following simple problem.

Example 2

A present sum of money (P) at an annual interest rate (I), if kept in a bank for N years would amount to a future sum (F) at the end of that time according to the equation $F = P(1+I)^N$. Prepare a flowchart for $P = \$100$ $I = 0.07$ $N = 5$ years and compute and output the values of F for all values of N from 1 to 5. Figure 9-2 is a flowchart for this situation.

Figure 9-1 Flowchart Symbols

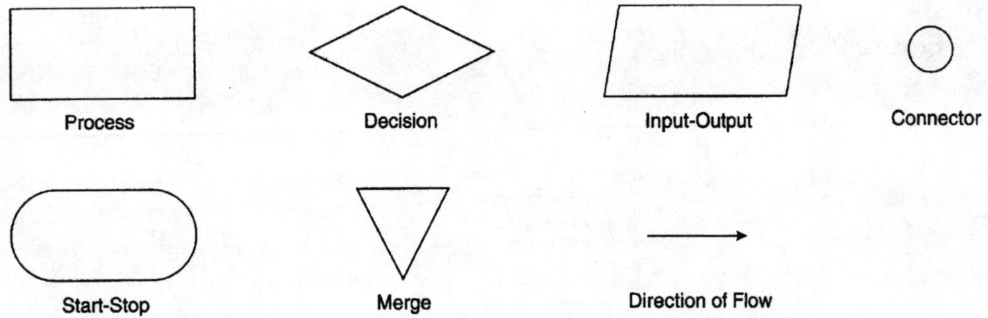

Figure 9-2

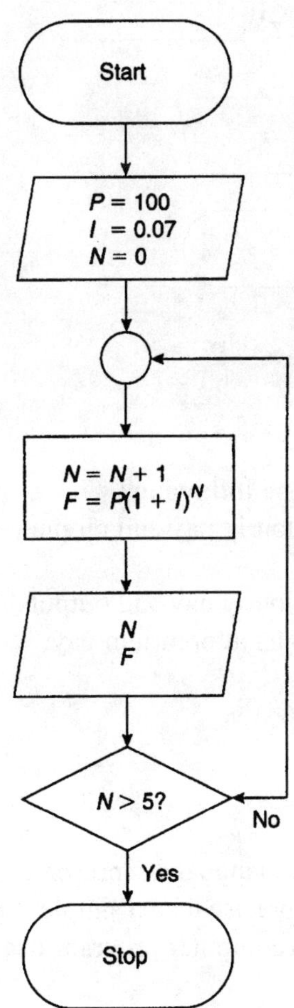

Example 3
Consider the flowchart in Figure 9-3.

Figure 9-3

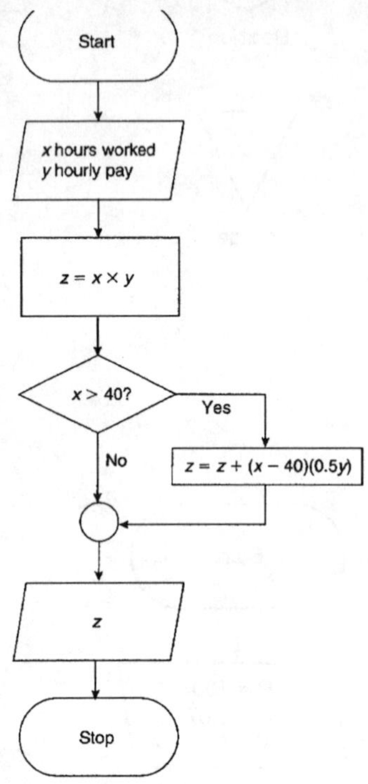

The computation does which of the following?

(a) Inputs hours worked and hourly pay and outputs the weekly paycheck for 40 hours or less.

(b) Inputs hours worked and hourly pay and outputs the weekly paycheck for all hours worked including over 40 hours at premium pay.

The answer is (b).

Pseudocode

Pseudocode is an English-like language representation of computer programming. It is carefully organized to be more precise than a simple statement, but may lack the detailed precision of a flowchart or of the computer program itself.

Example 4

Prepare pseudocode for the computer problem described in example 9-2.

Solution

```
INPUT P,I, and N = 0
DOWHILE  N < 5
COMPUTE  N = N + 1
         F = P ( 1 + I ) ^N
OUTPUT   N,F
IF N > 5 THEN ENDO
```

Spreadsheets

For today's engineers the ability to create and use spreadsheets is essential. The most popular spreadsheet programs are Microsoft's **Excel**, Novell's **Quattro Pro**, and Lotus **1-2-3**. Each of these programs use similar construction, methods, operators, and relative references.

Three type of information may be entered into a spreadsheet: text, values, or formulas. Text includes labels, headings and explanatory text. Values are numbers, times or dates. Formulas combine operators and values in an algebraic expression.

A **cell** is the intercept of a column and a row. Its location is based upon its column-row location, for example B3 would be the intercept of column B and row 3. Column labels are across the spreadsheet and row labels are on the side. To change a cell entry, the cell must be highlighted using either an address or pointer.

A group of cells may be called out by using a **range**. Cells A1, A2, A3, A4 could be called out using the range reference A1:A4 (or A1..A4). Similarly, A2,B2,C2,D2 would use the range reference A2:D2 (or A2..D2).

In order to call out a block of cells, a range call out might be A2:C4 (or A2..C4) and would reference the following cells:

A2 B2 C2
A3 B3 C3
A4 B4 C4

Formulas may include cell references, operators (such as +,-,*,/) and functions (such as SUM, AVG). The following formula SUM(A2:A6) or SUM(A2..A6) would be evaluated as equal to A2+A3+A4+A5+A6.

Relational References

Most spreadsheet references are relative to the cell's position. For example if the content of cell A5 contained B4 then the value of A5 is the value of the cell up one and over one. The relational reference is most frequently used in table tabulations as the following example for an inventory where cost times quantity equals value and the sum of the values yields the total inventory cost.

Inventory Valuation

	A	B	C
1 Item	Cost	Quantity	Value
2 box	5.2	2	10.4
3 tie	3.4	3	10.2
4 shoe	2.4	2	4.8
5 hat	1.0	1	1.0
6 Sum			26.4

In C2 the formula A2*B2
In C3 the formula A3*B3 and so forth.
For the Summation use the function SUM. In C6 use SUM(C2:C5)

Instead of typing in each cell's formula, the formula can be copied from the first cell to all of the subsequent cells by first highlighting cell C2 then dragging the mouse to include cell C5. The first active cell is C2 would be displayed in the edit window. Type the formula for C2 as A2*B2 and hold the control key down and the enter key is pressed. This copies the relational formula to each of the highlighted cells. Since the call is relational, then in cell C3 the formula is evaluated as A3*B3. In C4 the cell is evaluated as A4*B4. Similarly, edit operations to include copy, or fill operations simplify the duplication of relational formula from previously filled in cells.

Arithmetic Order of Operation

Operations in equations use the following sequence: exponentiation, multiplication and division, followed by addition and subtraction. Parentheses in formulas override normal operator order.

Absolute References

Sometimes a reference to a cell must be used that should not be changed, such as an data variable. An absolute reference can be named by inserting a dollar sign "$" before the column-row reference. If B2 is the data entry cell, then by using B2 as its reference in another cell, the call will always be evaluated to cell B2, even though it may be copied to another cell. Mixed reference can be made by using the dollar sign for only one of the elements of the reference. The reference B$2 is a mixed reference in that the row does not change but the column remains a relational reference.

The power of spreadsheets makes repetitive calculations very easy. In many operations, periods of time are normally new columns. Each element becomes a row item with changes in time becoming the columns.

All spreadsheet programs allow for changes in the appearance of the spreadsheet. Headings, borders, or type fonts are usual customizing tools.

NUMERICAL METHODS

This portion of numerical methods include techniques of finding roots of polynomials by Routh-Hurwitz criterion and Newton methods, Euler's techniques of numerical integration and the trapezoidal methods, and techniques of numerical solutions of differential equations.

Root Extraction

Routh-Hurwitz Method (Without Actual Numerical Results):

Root extraction, even for simple roots (i.e., without imaginary parts), can become quite tedious. Before attempting to find roots, one should first ascertain whether they are really needed or whether just knowing the area of location of these roots will suffice. If all that is needed is knowing whether the roots are all in the left half plane of the variable (such as in the s-plane when using Laplace transforms -- such as frequently the case in determining system stability in control systems), then one may use the Routh-Hurwitz criterion. This method is fast and easy even for higher ordered equations. As an example, consider the following polynomial:

$$P(x) = \prod_{m=1}^{n}(x-a_m) = x^n + a_1 x^{n-1} + a_2 x^{n-2} + \ldots + a_{n-1} \qquad (9\text{-}1)$$

Here, finding the roots for n>3 can become quite tedious without a computer; however if one only needs to know if any of the roots have positive real parts, then use the Routh-Hurwitz method. Here, an array may be made listing the coefficients of every other term starting with the highest power, n, on a line, then a following line may be made listing the coefficients of the terms left out of the first row. Following rows are constructed using Routh-Hurwitz techniques, and after completion of the array, one merely checks to see if all the signs are the same (unless there is a zero coefficient -- then something else needs to be done) in the first column; if none, no roots will exist in the right half plane. A simple technique used in control systems (for details, see almost any text dealing with stability of control systems. A short example follows:

$F(s) = s^3 + 3s^2 + 2s + 10$, $= (s+?)(s+?)(s+?)$

Array:

s^3	1	2
s^2	3	10
s^1	-4/3	0
s^0	10	0

Where the s^1 term is formed as: (3x2 -10x1)/3 = -4/3. For details refer to any text on control systems or numerical methods.

Here, there are two sign changes, one from 3 to -4/3, and one from -4/3 to 10; this means there will be two roots in the right half plane of the s-plane, which yields an unstable system; this technique represents a great savings in one's time without having to actually factor the equation.

Newton's Method

The use of Newton's method of solving a polynomial and the use of iterative methods can greatly simplify the problem. This method utilizes synthetic division and is based upon the remainder theorem. This synthetic division requires estimating a root at the start, and, of course, the best estimate is the actual root. The root is the correct one when the remainder is zero. (There are several ways of estimating this root, including a slight modification of the Routh-Hurwitz criterion.)

By taking a $P_n(x)$ polynomial (see equation 9-1) and dividing it by an estimated factor $(x-x_1)$, the result is a reduced polynomial of degree n-1, $Q_{n-1}(x)$, plus a constant remainder of b_{n+1}. Thus, another way of describing equation 9-1 is,

$$P_n(x)/(x-x_1) = Q_{n-1}(x) + b_{n-1} / (x-x_1) \quad \text{or as} \quad P_n(x) = (x-x_1)Q_{n-1}(x) + b_{n-1} \tag{9-2}$$

If one lets $x=x_1$, the equation 9-2 becomes,

$$P_n(x=x_1) = (0)Q_{n+1}(x) + b_{n+1} = b_{n+1}\,. \tag{9-3}$$

Equation 9-3 leads directly to the remainder theorem: "The remainder on division by $(x-x_1)$ is the value of the polynomial at $x= x_1$, $P_n(x_1)$." *

Newton's method (actually, the Newton-Raphson method) for finding the roots for an nth order polynomial is an iterative process involving obtaining an estimated value of a root (leading to a simple computer program). The key to the process is getting the first estimate of a possible root; without getting too involved, recall the coefficient of x^{n-1} represents the sum of all of the roots, and the last term represents the product of all *n* roots, then the first estimate can be "guessed" within a reasonable magnitude. After a first root is chosen, find the rate of change of the polynomial at the chosen value of the root to get the next closer value of the root, x_{n+1}. Thus the new root estimate is based on the last value chosen,

$$x_{n+1} = x_n - P(x_n)/P'(x_n), \quad \text{where } P'(x_n) = dP(x)/dx \text{ evaluated at } x=x_n \tag{9-4}$$

NUMERICAL INTEGRATION

Numerical integration routines are extremely useful in almost all simulation type programs, design of digital filters, theory of z-transforms, and almost any problem solution involving differential equations. And since digital computers have essentially replaced analog computers (which were almost true integration devices), the techniques of approximating integration are well developed. Several of the techniques are briefly reviewed below.

Euler's Method

For a simple first order differential equation, say $dx/dt + ax = af$, one could write the solution as a continuous integral or as an interval type one:

$$x(t) = \int^{t}[-ax(\tau)+af(\tau)]d\tau \tag{9-5a}$$

$$x(kT) = \int^{kT-T}[-ax+af]d\tau + \int_{kT-T}^{kT}[-ax+af]d\tau = x(kT-T) + A_{rect} \tag{9-5b}$$

Here, A_{rect} is the area of $(-ax+af)$ over the interval $(kT-T) < \tau < kT$. One now has a choice looking back over the rectangular area or looking forward. The rectangular width is, of course, T. For the forward looking case presented in a first approximation for x_1 is**,

$$x_1(kT) = x_1(kT-T) + T[ax1(kT-T)+af(kT-T)T = (1-aT)x_1(kT-T) + aTf(kT-T) \tag{9-5c}$$

*Gerald & Wheatley, *Applied Numerical Analysis*, Addison-Wesley, 3rd ed., 1985.

**This method is as presented in Franklin & Powell, *Digital Control of Dynamic Systems*, Addison-Wesley, 1980, page 55.

Or, in general, for Euler's forward rectangle method, the integral may be approximated in its simplest form (using the notation t_{K+1} - t_K for the width, instead of T which is kT-T) as,

$$\int_{t_K}^{t_{K+1}} x(\tau)d\tau \approx (t_{K+1}-t_K)x(t_K) \tag{9-6}$$

Trapezoidal Rule

This trapezoidal rule is based upon a straight line approximation between a function, f(t), at t_o and t_1. To find the area under the function, say a curve, is to evaluate the integral of the function between point a and b. The interval between these points are subdivided into subintervals; the area of each subinterval is approximated by a trapezoid between the end points. It will only be necessary to sum these individual trapezoids to get the whole area; by making the intervals all the same size, the solution will be simpler. For each interval of delta t (i.e., t_{K+1} - t_K), the area is then given by,

$$\boxed{\int_{t_K}^{t_{K+1}} x(\tau)d\tau \approx (1/2)(t_{K+1}-t_K)[x(t_{K+1})+x(t_K)]} \tag{9-7}$$

This equation gives good results if the delta t's are small but it is for only one interval and is called the "local error". This error may be shown to be -(1/12)(delta t)^{3}f''(t=ξ_1), where ξ_1 is between t_o and t_1. For a larger "global error" it may be shown that,

$$\text{Global error} = -(1/12)(\text{delta } t)^3\ [f''(\xi_1) + f''(\xi_2) + \ldots + f''(\xi_n)] \tag{9-8}$$

Following through on equation 9-8 allows one to predict the error for the trapezoidal integration; This technique is beyond the scope of this review or probably the examination; however for those interested, please refer to pages 249-250 of the previous mentioned reference to Gerald & Wheatley.

NUMERICAL SOLUTIONS OF DIFFERENTIAL EQUATIONS

This solution will be based upon a first order ordinary differential equations. However the method may be extended to higher ordered equations by converting them to a matrix of first ordered ones.

Integration routines produce values of system variables at specific points in time and update this information at each interval of delta time as T (delta $t = T = t_{k+1} - t$). Instead of a continuous function of time, $x(t)$, the variable x will be represented with discrete values such that $x(t)$ is represented by x_0, x_1, x_2, ..., x_n . Consider a simple differential equation as before, as (based upon Euler's method),

$dx/dt + ax = f(t)$.

Now assume the deta time periods, T, are fixed (not all routines use fixed step sizes), then one writes the continuous equation as a difference equation where: $dx/dt \approx (x_{k+1} - x_k)/T = -ax_k + f_k$ or, solving for the updated value, x_{k+1},

$$(x_{k+1}) = x_k - Tax_k + Tf_k \text{ for fixed increments.} \qquad (9\text{-}9a)$$

By knowing the first value of $x_{k=o}$ (or the initial condition), the solution may be achieved for as many "next values" of x_{k+1} as desired for some value of T. The difference equation may be programmed in almost any high level language on a digital computer; however, T must be small as compared to the shortest time constant of the equation (here, $1/a$).

The following equation (with the "*f*" term meaning a "function of" rather than as a "forcing function" term -- as used in equation 9-5a) is in a more general form of equation 9-9a. This equation is obtained by letting the notation (x_{k+1}) become $y[k+1)\Delta t]$ and is written (perhaps somewhat more confusing) as,

$$\boxed{y[(k+1)\Delta t]\ y(k\Delta t) + \Delta t f[(y(k\Delta t),\ k\,\Delta t]]} \qquad (9\text{-}9b)$$

Reduction of Differential Equation Order

To reduce the order of a linear time dependent differential equation, the following technique is used. For example, assume a second order one:

$x'' + ax' + bx = f(t)$, define $x = x_1$ and $x' = x_1' = x_2$, then, $x_2' + ax_2 + bx_1 = f(t)$,

$$x_1' = x_2 \qquad \text{<--- By definition.}$$
$$x_2' = -bx_1 + ax_2 + f(t)$$

This equation can be extended to higher order systems and, of course, be put in a matrix form (called the state variable form). And it can easily be set up as a matrix of first order difference equation for solving digitally.

Packaged Programs

Most currently available packaged simulation programs use algorithms not necessarily based upon Euler's methods but more advanced methods such as the Runge- Kutta method. Automatic variable step size methods like Milne's may also be used. However, as mentioned before, these routines are all built into the packaged programs and may be transparent to the user. The user of a specialized program may be without knowledge of the high level language being employed (except for certain modifications).

Glossary of Computer Terms

Accumulators	Registers which hold data, address, or instructions for further manipulations in the ALU
Address bus	Two way parallel path connecting processors and memory containing addresses
AI	Artificial Intelligence
Algorithm	A sequence of steps applied to a given data set which solve the intended problem
Alphanumeric data	Data containing the characters a,b,c...z,0,1,2,..9

ALU	Arithmetic and logic unit
ASCII	American Standard Code for Information Interchange ,7 bit/character (Pronounced AS-key)
Asynchronous	Form of communications which message data transfer is not synchronous with the basic transfer rate requiring start/stop protocol
Baud rate	Bits per second
BIOS	Basic input/output system
Bit	0 or 1
Buffer	Temporary storage device
Byte	8 bits
Cache Memory	Fast look-ahead memory connecting processors with memory offering faster access to often used data
Channel	Logic path for signals or data
CISC	Complex instruction-set control
Clock rate	cycles per second
Control bus	Separate physical path for control and status information
Control unit	Fetches, decodes instructions to control the operations of registers and ALU's
CPU	Central Processing Unit, primary processor
Data buffer	Temporary storage of data
Data bus	Separate physical path dedicated for data
Digital	Discrete level or valued quantification vs analog or continuous valued
Duplex Communication	Communications mode where data is transmitted in both directions, but only in one direction at any one time
Dynamic memory	Storage which must be continually hardware refreshed to remain
EBCDIC	Extended Binary Coded Decimal Interchange Code - 8 bits/character (Pronounced EB-see-dick)
EPROM	Erasable programmable read-only memory
Expert systems	Programs with AI which learn rules from external stimuli
Floppy disk	Removable disk media in various sizes, 5-¼", 3-½"
Flowchart	Graphical depiction of logic using shapes and lines
G-byte	Giga bytes 1,073,741,824 bytes or 2^{30}
Half-duplex communication	2-way communications path in which only 1 direction operates at a time (transmit or receive)
Handshaking	Communications protocol to start/stop data transfer
Hard disk	Disk which has non-removable media
Hardware	Physical elements of a system
Hexadecimal	Numbering system (base 16) uses 0-9,A,B,...F
Hierarchical database	Database organization containing hierarchy of indexes/keys to records
I/O	Input/Output devices such as terminal, keyboard, mouse, printer

IR	Instruction register
K-bytes	Kilo-bytes 1024 bytes or 2^{10}
LAN	Local Area Network
LIFO	Last In-First Out
LSI	Large scale integration
Main memory	That memory seen by CPU
M-bytes	Mega-bytes 1,048,5 76 bytes or 2^{20}
Memory	Generic term for random access storage
Microprocessor	Computer architecture with Central Processing Unit in one LSI chip
MODEM	Modulator-demodulator
MOS	Metal Oxide Semiconductor
Multiplexer	Device which switches several input sources one at a time to an output
Narrow band	Frequency domain reference to channel band width versus baseband
Nibbles	4 bits
Non-volatile memory	As opposed to volatile memory, does not need power to retain present state
Number systems	Method of representing computer data into human readable information
OCR	Optical Character Recognition
OS	Operating system
OS memory	Memory dedicated to the OS, not useable for other functions
Parallel interface	A character(8 bit) or word(16 bit) interface with as many wires as bits in interface plus data clock wire.
Parity	Method for detecting errors in data, 1 extra bit carried with data, even or odd, to make the sum of one bits in a data stream
PC	Program counter, or personal computer
Peripheral devices	Input/Output devices not contained in Main processing hardware
Program	A sequence of computer instructions
PROM	Programmable Read-Only Memory
Protocols	Established set of handshaking rules enabling communications
Pseudocode	An English-like way of representing structured programming control structures
RAM	Random Access Memory
Real time/Batch	Method of program execution; Real-time implies immediate execution, Batch mode is postponed until run on a group of related activities
Relational database	Database organization which related individual elements to each other without fixed hierarchical relationships
RISC	Reduced instruction set computer
ROM	Read Only Memory
Scratchpad memory	Highspeed memory either in hardware or software
Sequential Storage	Memory (usually tape) accessed only in sequential order (n, n+1,..)

Serial Interface	Single data stream which encodes data by individual bit per clock period
Simplex communication	One-way communications
Software	Programmable logic
Stacks	Hardware memory organization implementing Last In-Last Out access
Static memory	Memory which does not require intermediate refresh cycles to retain state
Structured Programming	Use of Structured programming constructs such as Do-While, If-Then, Else
Synchronous	Communications mode which data and clock are at same rate
Transmission speed	Rate at which data is moved in baud. (bits per second,bps)
Virtual memory	Addressible memory outside physical address bus limits through use of memory mapped pages
Volatile memory	Memory whose contints are lost when power is removed.
VRAM	Video memory
Wide-band	300-3300 Hz
Words	8,16,or 32 bits
WYSIWYG	What you see is what you get
16-bit	Basic organization of data with 2-bytes per word
32-bit	Basic organization of data with 4-bytes per word
64-bit	Basic organization of data with 8-bytes per word
80386	Intel's microprocessor architecture based on 16 data address bus, extended virtual memory, external math coprocessor
80486	Upgrade to 80386 incorporating Math coprocessor within VLSI
80586	Intels microprocessor architecture based on 16 bit data, 32 bit address bus

Problems and Solutions

9-1. In spreadsheets, what is the easier way to write
B1 + B2 + B3 + B4 + B5
(a) Sum(B1:B5)
(b) (B1..B5)Sum
(c) @B1..B5SUM
(d) @SUMB2..B5

9-2. The address of the cell located at column 23 and row C is
(a) 23C
(b) C23
(c) C.23
(d) 23.C

9-3 Which of the following is not correct?

(a) A CD-ROM may not be written to by a PC.
(b) Data stored in a batch processing mode is always up-to-date.
(c) The time needed to access data on a disk drive is the sum of the seek time, the head switch time, the rotational delay, and the data transfer time.
(d) The methods for storing files of data in secondary storage are: sequential file organization, direct file organization, and indexed file organization.

9-4. Which of the following is false?
(a) Flowcharts use symbols to represent input/output, decision branches, process statements and other operations.
(b) Pseudocode is an English-like description of a program.
(c) Pseudocode used symbols to represent steps in a program.
(d) Structured programming breaks a program into logical steps or calls to subprograms.

9-5. In Pseudocode using DOWHILE , the following is true:
(a) DOWHILE is normally used for decision branching.
(b) The DOWHILE test condition must be false to continue the loop.
(c) The DOWHILE test condition tests at the beginning of the loop.
(d) The DOWHILE test condition tests at the end of the loop.

9-6. A Spreadsheet contains the following formulas in the cells:

	A	B	C
1		A1+1	B1+1
2	A1^2	B1^2	C1^2
3	Sum(A1:A2)	Sum(B1:B2	Sum(C1:C2)

If 2 is placed in cell A1, what is the value in Cell C3?

(a) 12
(b) 20
(c) 8
(d) 28

9-7. A matrix contains the following:

	A	B	C	D
1		3	4	5
2	2	A$2		
3	4			
4	6			

If you copy the formulas from B2 into D4, what is the equivalent formula in D4?

(a) A$2
(b) C4
(c) C4
(d) C$2

9-8. A Processing system is processor limiter when performing sorting of small tables in memory. Which would speed up computations?

(a) Adding more main memory
(b) Adding virtual memory
(c) Adding cache memory
(d) Adding peripheral memory

9-9. A small PC processing system is performing large (1MB) matrix operations which is currently I/O limited since memory is limited to 1Mbyte. Which would speed up computations?

(a) Adding more main memory
(b) Adding virtual memory
(c) Adding cache memory
(d) Adding peripheral memory

9-10. Transmission Protocol: Serial, asynchronous, 8 bit ASCII, 1 Start, 1 Stop, 1 Parity bit, 9600 bps. How long will it take for a 1Kbyte file to be transmitted through the link?

(a) 0.85 sec.
(b) 0.96 sec.
(c) 1.07 sec.
(d) 1.17 sec.

9-11. Transmission Protocol: Serial, synchronous, 8 bit ASCII, 1 Parity bit, 9600 bps. How long will it take for a 1Kbyte file to be transmitted through the link?

(a) 0.85 sec.
(b) 0.96 sec.
(c) 1.07 sec.
(d) 1.17 sec.

9-12. The binary representation 10101 corresponds to which base 10 number?

(A) 3
(B) 16
(C) 21
(D) 10,101

9-13. For the base 10 number of 30, the equivalent binary number is

(A) 1110
(B) 0111
(C) 1111
(D) 11110

9-14. On personal computers the data storage device with the largest storage capacity is most likely to be

(A) 3½" diskette
(B) hard disk
(C) random access memory
(D) 3½" HD diskette

9-15.

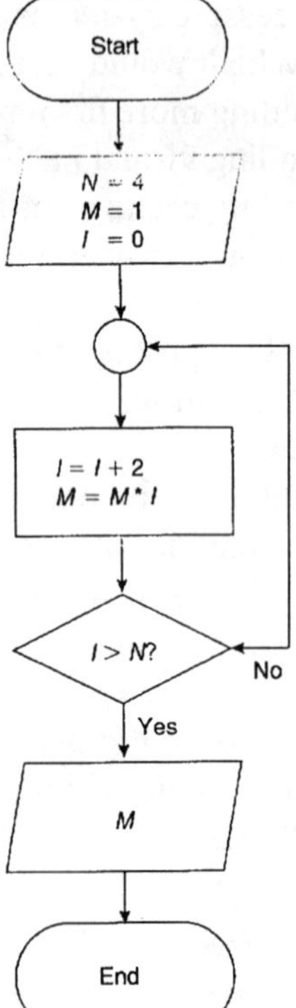

Figure 9-4.

The output value of M in Figure 9-4 is closest to

(A) 1
(B) 2
(C) 8
(D) 48

9-16. Pseudocode can best be described as

(A) A simple letter-substitution method of encryption.
(B) An English-like language representation of computer programming.
(C) A relational operator in a database.
(D) The way data are stored on a diskette.

9-17.

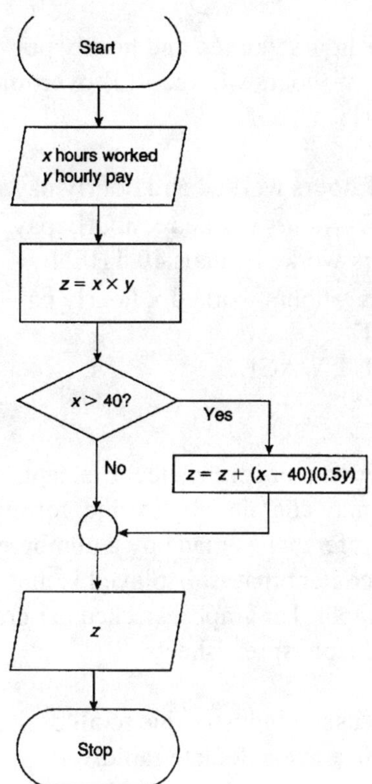

Figure 9-5

9-18

The pseudocode that best represents Figure 9-5 is:

(A) INPUT X,Y
WAGE = X x Y
IF X less than 40 THEN
WAGE = WAGE + OVERTIME PAY
END IF
OUTPUT WAGE

(B) INPUT hours worked and hourly pay
WAGE = hours worked x hourly pay
IF hours worked greater than 40 THEN
WAGE = WAGE + (hours worked -40)(overtime wage supplement)
END IF
OUTPUT WAGE

(C) INPUT hours worked and hourly pay
WAGE = (hours worked -40)(overtime rate) + 40(hourly pay)
OUTPUT WAGE

(D) PRINT hours worked and hourly pay
WAGE = hours worked x hourly pay
IF hours worked equals 40 THEN
WAGE = hours worked x hourly pay + (hours worked -40) x overtime pay
END IF
OUTPUT WAGE

9-18.
All of the following is true of spreadsheets except:

(A) A cell may contain label, value, formula or function
(B) A cell reference is made by a numbered column and lettered row reference
(C) A cell content that is displayed is the result of a formula entered in that cell
(D) Line graphs, bar graphs, stacked bar graphs and pie charts are typical graphs created from spreadsheets.

9-19.
Computers find wide use in industry due to all of the following reasons except one:

a) They are fast and calculate rapidly.
b) They are accurate.
c) They can react quickly to unanticipated situations.
d) They can store large amounts of information.

Solutions

SOLN 9- 1

Sum(B1:B5) or @Sum(B1..B5) Answer is (a).

SOLN 9-2.
Answer is (b).

SOLN 9-3.
Batch mode processing always have delays in updating the database. Answer is (b).

SOLN 9-4.
Pseudocode does not use symbols but uses English-like statements such as IF-THEN, and DOWHILE. Answer is (c).

SOLN 9-5.
IFTHEN is normally used for branching. The DOWHILE test condition must be true to continue branching and the test is done at the beginning of the loop. The DOUNTIL test is done at the end of the loop. Answer is (c).

SOLN 9-6.
Plugging 2 into spreadsheet produces the following matrix:

	A	B	C
1	2	3	4
2	4	9	16
3	6	12	20

The value of C# is 20. Answer is (b).

SOLN 9-7.
The formula contains mixed references. The "$" implies absolute row reference while the column is relative. The result of any copy would eliminate any answer except for absolute row 2 entry relative column reference "A" gets replaced by "C". The cell contains C$2. The answer is (d).

SOLN 9-8.
Processor limited sorting on small tables suggest either speeding up processor cycles or providing faster memory. Since speeding up clock is not an option, then making memory faster is the answer. Adding more memory, or virtual memory does nothing for small tables. Only adding cache memory would allow the CPU to fetch recently used data without full memory cycles thereby speeding up the sorting process. Answer is (c).

SOLN 9-9.
Processing is I/O limited since all of matrix can not fit into memory. Since this is a large matrix, cache memory probably would not effect processing. The best solution is adding more main memory to fit this problem entirly in memory. Answer is (a).

SOLN 9-10.

1 Kbyte=2^{10} bytes=1024

1024 bytes + 3 overhead bits/byte = 1024(8 bits/byte +3 bits/byte overhead)
= 11,264 bits

Minimum transmission time = 11264 bits/9600 = 1.17 sec. Answer is (d).

SOLN 9-11.

1 Kbyte=2^{10} bytes=1024

1024 bytes + 1 overhead bit/byte = 1024(8 bits/byte + 1 bit/byte overhead) = 9216 bits

Minimum transmission time = 9216 bits/9600 = 0.96 sec. Answer is (b).

SOLN-9-12

The binary (base 2) number system representation is

	2^4	2^3	2^2	2^1	2^0	
	16	8	4	2	1	
Binary number	1	0	1	0	1	
	___	___	___	___	___	
	16	0	4	0	1	= 21

Answer is (C).

SOLN-9-13.

2^5	2^4	2^3	2^2	2^1	2^0	
32	16	8	4	2	1	
	1	1	1	1	0	= 30

Answer is (D).

SOLN-9-14. Answer is (B).

SOLN-9-15. Answer is (D).

SOLN-9-16. Answer is (B).

SOLN- 9-17. Answer is (B).

SOLN-9-18.

A cell reference is made by a lettered column and a numbered row, example C3. Answer is (B).

SOLN-9-19.

Computers are fast and accurate and can store large amounts of information, but they cannot "think". A computer is actually a "great stupid beast" and can only do what it is programmed to do, so it cannot react to a situation for which it has not been programmed. It cannot react to an unanticipated situation. The correct answer is c).

Chapter 10
Energy Conversion and Power Plants

The conversion of energy to useful power constitutes a large portion of mechanical engineering endeavor. Energy is available in many forms and includes gas, oil, coal, peat and other forms of organic energy as well as energy from water, the sun, nuclear reactors, and underground thermal sources. The major types of conversion systems include power plants to produce electricity, turbines, reciprocating steam engines, and internal combustion engines.

Perhaps the major portion of energy conversion takes place in the many large power plants which produce electricity. The final step in the production of electricity is achieved by means of generators, however, the generators must be driven by prime movers. The overall efficiency of a power plant equals the power output in watts divided by the heat input in joules.

EXAMPLE 10-1

A small company power plant produces 54 kW of useful electric power. The generator is 96% efficient and a pump to recirculate the working substance uses 2.5 kW. A heat source supplies 10.2 MJ/min. (a) What is the power output of the heat engine, (b) what is the overall thermal efficiency of the power plant, and (c) what is the amount of heat rejected?

Solution

Assume that power from the generator is used to operate the recirculating pump. The 2.5 kW must then be added to the 54 kW of useful power to give a total generator output of 56.5 kW.

56.5/0.96 = 58.85 kW power to the generator which is the work power output of the heat engine (a)

10.2 MJ/min = 170 kJ/s or 170 kW.

The overall thermal efficiency of the power plant equals-- (thermal output)/(thermal input)

Thermal efficiency = 54.0/170 = 0.3176 or 31.8% (b)

Heat rejected = (1.0 - 0.318) × 10.2 = 6.96 MJ/min (c)

EXAMPLE 10-2

A 300 kW turbine is to operate at 3,600 rpm with an inlet condition of 2.00 MPa steam at 100 C superheat and discharge at 70 kPa. How many kg of steam per hour will it take to operate the turbine?

Solution

A preliminary estimate is all that is required.

The pressures are given as gage pressures, so it is necessary to add one atmosphere, 101.3 kPa, to obtain absolute pressures. From the steam tables the initial enthalpy is estimated to be 3.057 MJ/kg for a pressure of 2.101 MPaa. The entropy is 6.802 J/kg-○K. Following down a constant entropy line on a Molier diagram gives a value of 2.477 MJ/kg for the enthalpy at discharge. The steam will give up 0.580 MJ/kg of steam.

300 kW = 300 kJ/s or 1,080 MJ/h If the turbine operated at 100% efficiency it would require--

1,080/0.580 = 1,862 kg/h

It is estimated that a turbine of this size operating at the given temperature would have an efficiency of about 60%, so the steam requirement would be about 3,100 kg/h.

SOLAR ENERGY
An engineer is planning to build a vehicle to be powered by solar energy. His solar panels are rated to provide electricity in the amount of 10% of incident energy. The radiant solar energy is estimated at 0.140kW/m². It has been determined that 1.00 kW will be needed to satisfy the requirements of the proposed vehicle. What area of solar panels will be required?

Each m² of solar panel will produce 140 W

An area of 1.00/0.140 = 7.143 m² will be needed.

PRONY BRAKE A prony brake is sometimes used to measure the power output of an engine. A diesel engine was tested with a prony brake of the type shown in Fig. 10-1. For this setup L = 700 mm the tare load is 180 N, and the engine operated at a speed of 1,150 rpm at a certain throttle setting. What was the power output of the engine at this throttle setting if the force F = 2,200 N?

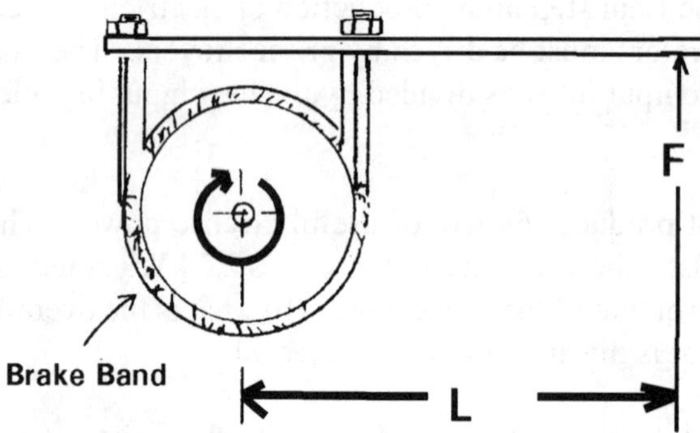

Figure 10-1. Prony Brake

The measured torque is 0.700 × (2,200 - 180) = 1,414 J
The power output = 2π x (1,150/60) × 1,414 = 170,285 W or 170.3 kW

OTTO CYCLE A gasoline engine operates on the air-standard Otto cycle. The efficiency can be determined from the equation given in the FE Handbook, $\eta = 1 - r^{1-k}$

where: r = compression ratio.

How much gasoline would be required to operate a tractor with a 75 kW engine for an 8-hour shift if the engine has a compression ratio of 8:1 and the gasoline has a heating value of 48.27 MJ/kg? The specific gravity of gasoline is 0.739. The tractor is to operate at full power throughout the shift.

The volume would probably best be determined in liters. A liter of water has a mass of 1.00 kg, so one liter of gasoline would have a mass of 0.739 kg giving a heating value of 35.67 MJ/ℓ. One kW equals 1,000 J/s or 3.60 MJ/h, so a power output of 75 kW for 8.00 hours would require 2.160 GJ. 2,160/35.67 = 60.6 liters if the conversion of energy were 100 efficient. The estimated efficiency of the air-standard Otto cycle engine equals $\eta = 1 - 8.0^{-0.4}$ so
η = 56.47 efficient, so the total fuel requirement for an
8-hour shift would be 107.3 liters.

INDICATOR CARDS The power output of a reciprocating engine can be measured with the aid of an engine indicator. This is a device which measures pressure in the cylinder vs. travel of the piston. Fig. 10-2 shows a sample diagram for a four-stroke-cycle gas engine. Note the intake portion of the cycle at slightly below atmospheric pressure from points (1) to (2). Compression from (2) to (3), fuel burning at the top of the stroke and expansion from (3) to (4) with exhaust from (4) to (1). The area under the P-V curve is measured with a planimeter and divided by the stroke length to give an average pressure, or the M.E.P., the mean effective pressure. This, times the area of the piston, times the stroke length gives the work done per piston power stroke.

EXAMPLE 10-3

A four-cylinder, four-cycle engine with 12.00 cm diameter pistons and an 18 cm stroke operating at a speed of 500 rpm, gave an indicator diagram as shown in Fig. 10-2. The area under the curve (P-V diagram) equals 10.42 cm². The length of the diagram is 8.23 cm, and the spring constant of the indicator spring is 550 kPa/cm. (a) What is the M.E.P. and (b) What is the indicated power?

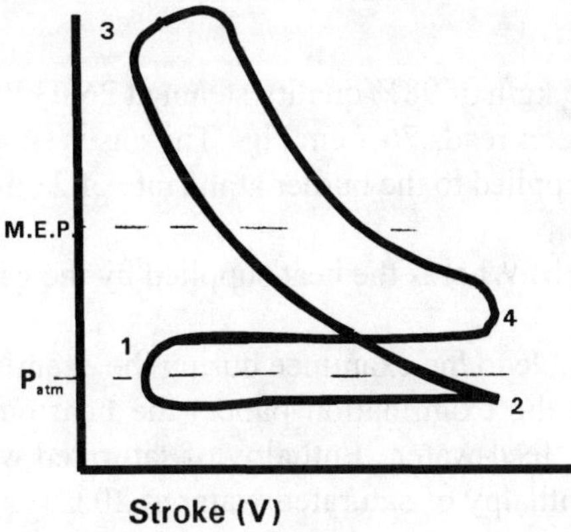

Figure 10-2. Otto Cycle

Solution

The average height equals 10.42/8.23 = 1.266 cm

Average pressure = 1.266 × 550 = 696.3 kPa (a)

The volume swept out by the piston on a power stroke

vol = 113.1 × 18 = 2,035.8 cm³ = 0.002036 m³

Work per power stroke = 696300 × 0.002036 = 1,418 J

In a four-cycle engine each piston makes a power stroke every two revolutions, so the engine would make:

500/2 × 4 = 1,000 power strokes per minute

Indicated power output of engine = 1,418 kJ/m

1,418/60 = 23.63 kJ/s or 23.63 kW

EXAMPLE 10-4

A large stationary diesel engine produces 1,500 kW to run an electric power plant and has a specific fuel rating of 0.255 kg/kW-hr. The fuel has a heating value of 42.00 MJ/kg. The air fuel ratio is 16:1. The engine exhaust temperature is 430 C and the ambient conditions are 30 C and one standard atmosphere. What flow rate of cooling water is necessary if the engine cooling apparatus permits a 15 C rise in water temperature through the engine.

Solution

The engine produces 1,500 kW of work or 5.40 GJ/h

Fuel consumption is 1,500 × 0.255 = 382.5 kg/h

or 16.065 GJ/h energy supplied to the engine. 33.61 % efficiency.

For every kg of fuel 16 kg of air would enter the engine. This mass would be heated from ambient temperature to exhaust temperature, an increase of 400 C. The specific heat of the mixture at the engine conditions can be taken as 1.089 kJ/kg-C. The heat absorbed by the air-fuel mixture equals $17 \times 382.5 \times 400 \times 1.089 = 2.832$ MJ/h = 0.0028 GJ
Estimate that 5.00% of the theoretical energy supplied by the fuel is lost due to water in the fuel, incomplete combustion, and dissipation to the surroundings.
The heat balance gives:

$16.065 - 0.8033 - 5.40 - .003 = 7.030$ GJ/h to be absorbed by the cooling system.

The specific heat of water is 4.18 kJ/kg-C from the FE Handbook so each kg of cooling water flow would absorb $4.18 \times 15 = 62.7$ kJ. Required water flow--

$9.859 \times 10^9/62,700 = 157.2$ Mg/h or 157.2 m^3/h

Q = 2,620 ℓ/m

STEAM BOILER

A steam boiler heated by gas delivers 115 kg/h of 98% quality steam at 26.0 kPa pressure. The feed-water temperature is 20 C and the barometer reads 76.7 cm Hg. The gas is supplied to the burner at 27.0 C and 10.0 cm water gage. Gas is supplied to the burner at the rate of 23 m^3/h. The fuel is rated at 20.1 MJ/m^3 at 20 C and 76 cm Hg.
(a) What is the heat output of the boiler, (b) What is the heat supplied by the gas, and (c) what is the efficiency of the boiler?

From the steam tables (not available to the examinee during the examination, but assuming pertinent values would be supplied on the examination paper) the heat output equals the heat contained in the steam less the heat in the feed-water. Enthalpy of saturated water at 26.0 kPa plus 98% of the heat of evaporation minus enthalpy of saturated water at 20 C.

$546 + 0.98 \times 2,175 - 84 = 2,594$ kJ/kg

Heat output of boiler = $2,594 \times 115 = 298.3$ MJ/h (a)

The gas is supplied to the boiler at different conditions than those at which it was rated, so it is necessary to convert the volume of gas supplied to the rated conditions. The gas is supplied at 27 C and a pressure of

$76.7 + 10/13.6 = 77.44$ cm Hg so the volume will be less than the standard volume measured at 76 cm Hg, and the temperature is higher so this will act to provide a higher volume than the standard volume. To correct the volume to the standard conditions--

Vol = $23.0 \times (77.44/76) \times [(273 + 20)/(273 + 27)]$

corrected volume = 22.89 m^3/h of standard gas

Heat supplied to boiler--

$22.89 \times 20.1 = 460.09$ MJ/h (b)

Boiler efficiency = 298.3/460.09 = 0.6484 or 64.8% (c)

FEED-WATER HEAT

It is sometimes desired to pre-heat feed-water using exhaust steam. The exhaust steam mixes with the new feed-water supply as it heats it. The final volume of feed-water then equals the amount heated plus the condensed steam used to heat the new supply.

EXAMPLE 10-5

Steam exhausted from a steam engine is used to heat feed-water in an open-type feed-water heater. How much steam is needed to raise the temperature of 500 kg of feed-water from 20 C to 95 C if the steam is exhausted from the engine at 7.00 kPa with a moisture content of 14 percent?

Solution

From a steam table the enthalpy of water at 20 C equals-- 84 kJ/kg and at 95 C equals 396 kJ/kg so the total amount of heat in the feed-water after heating would equal 500 × 396 = 198 MJ

Let m = mass of steam exhausted from engine

The feed-water makeup would contain (500 - m) × 84 kJ of heat. Steam at exhaust condition of (101.3 + 7.0) kPaa and 86% quality would contain

428 + 0.86 × 2,250 = 2,364 kJ/kg

The total heat contained in the feed-water at the final condition would equal 500 × 396 = 198 MJ

Heat balance gives 198,000 = (500 - m) × 84 + m × 2,364

or m = 156,000/2,280 = 68.4 kg of exhaust steam.

PROBLEMS

10-1. A large paper plant uses its own co-generation power plant to generate its electricity because:

a) It permits the use of a more efficient generator.
b) The company can better control the voltage.
c) It does away with high-power lines leading to the plant.
d) It provides a more efficient conversion of fuel.

10-2. A large municipal power plant produces 1,000 MW of power. The overall plant efficiency is 32.0%. Coal with a heating value of 24 MJ/kg is used to heat the boilers. How much coal is burned each day?

a) 10,500 metric tons
b) 11,250 metric tons
c) 12,000 metric tons
d) 12,500 metric tons

10-3. If the above 1,000 MW power plant were operating on nuclear fuel, how much fuel would be burned each day?

a) 15 grams
b) 75 grams
c) 22 grams
d) 3 grams

SOLUTIONS

10-1. An in-plant co-generation unit in a paper plant uses a "topping turbine", i.e. it takes the "top" off the steam supplied to the turbine and uses the heat that is ordinarily dumped in the condenser to dry the paper it produces. The result is a rather considerable increase in the overall efficiency of the fuel conversion.
The correct answer is d).

10-2. The heat requirement is 1,000/0.32 = 3,125 MJ/s.
The coal required equals 3,125/24.0 = 130.2 kg/s
11,250,000 kg/day or 11,250 metric tons/day ans.
The correct answer is b)

10-3.
The nuclear energy mass/energy equivalent is given by the relationship $E = mc^2$
$c = 299{,}792{,}000$ m/s (From the FE Handbook)
The energy requirement for the plant is 3,125 MJ/s
$E = \text{kg} \times 8.988 \times 10^{16}$ m²/s² or 8.988×10^{16} J/kg
The plant power requirement equals--
$3{,}125 \times 10^{6} \times 3{,}600 \times 24 = 270 \times 10^{12}$ J/day
$(270 \times 10^{12})/(8.988 \times 10^{16}) = 0.00300$ kg
or 3.00 grams of nuclear fuel
The correct answer is d)

Chapter 11
Automatic Control

Industrial and domestic processes are becoming more and more dependent upon mechanical control and less and less on human monitoring. In most cases automatic control is more reliable, more accurate, and faster than human regulation of a manufacturing process. It is also less expensive and companies have been forced to automate, and "robotize", to remain competitive. In addition, automatic controls as a rule result in increased personnel safety. There are two general types of automatic control--open-loop control, and closed-loop feedback control.

OPEN-LOOP CONTROL In this type of a control system there is no feedback. An example is a home owner's greenhouse system. The lights go on at a certain time as set by the gardener regardless of the amount of outside sunshine, and the watering system waters the interior plants regardless of the amount of soil moisture. The functions are preset and are not affected by the conditions they control. This type of system is little used in industry today, even plant and city outdoor lighting are often controlled by light-operated switches which only activate the lighting circuits when the natural sunlight is inadequate for expected activities.

CLOSED-LOOP FEEDBACK CONTROL This type of control is the most common type of automatic control used commercially and also domestically. It can be employed with any process in which the process variable to be controlled is measurable. A common system is a home heating system.

TWO-POSITION CONTROL This is the simplest of all control systems. The output of the controller is either on or off, 100% or 0%. In a domestic heating system when the temperature drops below a certain set point, an impulse is sent to the heating device and it goes on. The heating system stays on until the temperature reaches another set point and then the heating system is turned off. Since there is a residual capacity in the heating system, heat remaining in the furnace, in the ducts, and in the hot supply air, the temperature in the controlled space can continue to rise a certain amount before it stabilizes. To overcome what may be an undesirable variation in the controlled temperature a proportional control system can be employed rather than the on-off system described. In addition most systems are supplied with fan-limit switches which delay the start of the fan or blower until the temperature of the heat-supply system has reached a pre-determined value to prevent any uncomfortable blast of cold air when the system first turns on. And there is also a delay incorporated into the fan control system to maintain air flow after the furnace (heat supply) is shut off to prevent overheating of that unit due to retained heat. A domestic heating system has many automatic control mechanisms.

PROPORTIONAL CONTROL Proportional control for a large-volume heat supply system adjusts the amount of heat supplied to the amount required to perform the desired heating. For example, if the temperature is well below the desired point, the heater would be wide open to supply the maximum amount of heat of which it was capable. As the temperature nears the desired point, the heat supply would be gradually throttled back until it was shut off entirely when the controlled temperature reached the desired point.

EXAMPLE An office complex is heated by a furnace in the basement of the building which supplies heat to the entire building. In the winter the temperature is controlled by means of a thermostat on the office wall. When the temperature in the office falls below the set point, hot, furnace-heated air is mixed with the air supplied to the office by means of a control valve connecting the office air system to the hot air supply from the furnace. At the start of the heating-up process the control valve will open wide and relatively warm air will be supplied to the cool office complex. As the office air nears the desired temperature the control valve gradually closes until the temperature of the supply air is the same as the desired temperature of the office air.

EXAMPLE 11-1. A large warehouse has an office for the clerks at one end. A single thermostat controls the heat supply to the building. The warehouseman at the far end of the building consistently turns up the thermostat to give a comfortable temperature at the far end, while the clerks in the office at the other end roast before the entire barn-like building warms up. How would you solve this problem?
SOLUTION One way to handle the problem would be to provide two separate air supply points and two control locations. From the main heated-air supply duct attach a smaller duct to the office area. Control the flow of hot air through this off-shoot separately by means of a variable flow valve controlled by a thermostat in the office area. The large building would require much more heat than the smaller office area, so heat would always be available for the clerks.

REFRIGERATION SYSTEMS An important part of every refrigeration system is the automatic control system. An automatic control system eliminates the need for frequently adjusting manually operated valves to control the amount of cooling. It also provides many safety features which are especially important in our society's present-day emphasis on personnel safety.

The refrigeration load is primarily controlled by adjusting the expansion valve to modify the flow rate of refrigerant into the expansion coils. For a cooling system with an essentially constant load, such as a deep freeze storage facility, there will be little need for any variation in the rate of flow of refrigerant liquid into the cooling coils. In such an installation there is no requirement for constant monitoring and variation of the flow of refrigerant, and a manually-adjusted expansion valve is sufficient, and is still used in many plants.

In the case of comfort air conditioning, however, the situation is quite different. An extreme case is that of a restaurant. Early in the morning there will be a moderate cooling requirement when breakfast is served and the room is filled with customers even though the outside temperature is mild. When the crowd leaves the requirement will diminish but it will rise to a higher level during the noon-time rush. Then it will drop again only to rise once more during the dinner hour. The cooling load will vary due to the number of customers and also due to the weather. And it will fluctuate from hour to hour and from day to day in ways that cannot be predicted with any accuracy. The necessary refrigeration requirement thus cannot be predicted or programmed with any degree of accuracy. Manual control would require constant monitoring. In addition there is always the danger that the momentary refrigeration requirement would be over-estimated and too much liquid supplied to the cooling coils. If more refrigerant is supplied than can be used a slug can pass to the compressor and cause repeated lifting of the compressor head with accompanying racket. For such fluctuating load conditions automatic control of the flow of refrigerant into the cooling coils is a necessity.

Automatic control is used many other places such as inspection systems, automobile traffic control, materials handling, food dispensing for animals, and almost every place where a product or service must be adjusted to meet a demand.

EXAMPLE 11-2. An automatic production line is used to fill boxes. Occasionally a glitch occurs in the box printing line. The boxes are filled at the rate of 100/min. The boxes are ten cm wide and are spaced two cm apart. How rapidly must an inspection device respond to catch a defective box if it must scan a five cm distance to catch an unprinted box? The inspection device is actuated by an electric eye which alerts it when a new box is in front of it.

SOLUTION The distance from center line to center line of the boxes is 12 cm. The speed of the production line equals-- $100 \times 12 = 1{,}200$ cm/min The production line will travel five cm in $5/20 = 0.25$ sec

PROBLEMS

11-1. A product is held to mass limits of 1.5 kg plus or minus 10 mg. What must be the accuracy of the mass-measuring device if a tolerance of 1.00 mg is to be achieved?

a) 99.99%
b) 99.97%
c) 99.95%
d) 99.93%

11-2. An inspection belt line operates at a speed of 10 m/min. The objects being inspected are 15 cm long in the direction of travel and are spaced 30 cm apart. The automatic inspection device operates at a speed of 30 msec. How far ahead of the point of inspection should a mechanical pusher ram be placed to push a defective part off the belt if it takes the ram 500 msec to strike a defective part?

a) 12 cm
b) 9 cm
c) 6 cm
d) 14 cm

11-3. A pet owner has an automatic feeder which issues a measured amount of feed at a set time each day. This is an example of which of the following types of automatic control?

a) Feed back control.
b) Closed loop control.
c) Open loop control.
d) Automatic control.

SOLUTIONS

11-1. Applying the maximum error to the maximum reading gives that the scale must be accurate to one mg in 1.510 kg, or one part in 1,510. The accuracy must be--
1,509/1,510 = 99.934%
or the error must not exceed 0.066% of the maximum reading.
The correct answer is d).

11-2. The time for the ejection ram to strike a defective part is 530 msec. During that length of time the inspection belt would move--

$$\Delta L = 0.530 \times 1{,}000/60 = 8.833 \text{ cm}$$

The correct answer is b).

11-3. There is no feed back, the feeder issues a set amount of feed at the set time each day regardless of whether the pet is hungry or not. The control system is thus defined as an open-loop control system.
The correct answer is c).

Chapter 12

Refrigeration and HVAC

Refrigeration is a process whereby heat is extracted from a system. This was formerly accomplished by using ice which absorbed heat by melting, using its heat of fusion to absorb heat from the cooled substance. This method has largely been replaced by mechanical refrigeration which extracts heat from a cold reservoir and dumps it to a warmer one. The influence of ice refrigeration is still with us, however, in that refrigeration capacity is measured in tons, where a "ton of refrigeration" is the amount of refrigeration required to freeze one ton of water in 24 hours. In the U.S. system this equals--

2,000 lb × 144 Btu/lb = 288,000 Btu/(24-hour day) or 12,000 Btu/hr

This becomes 12.66 MJ/hr or 3,517 J/s or 3.517 kW, so one ton of refrigeration equals 3.517 kW.

MECHANICAL REFRIGERATION

Mechanical refrigeration is based on the principle that a liquid will vaporize at some specific pressure and when it vaporizes it will absorb the heat of vaporization. A mechanical refrigeration system consists of (1) an evaporator where a liquid refrigerant evaporates at a cold temperature (and a relatively low pressure) absorbing heat from the cooled mass, (2) a compressor which compresses the cold gas from the evaporator, (3) a condenser where the hot compressed gas is cooled to its liquid temperature, and (4) an expansion valve where the liquid expands to the lower pressure of the evaporator and a portion of the liquid vaporizes and cools the remaining portion to the refrigerating temperature as it flows into the evaporator.

ABSORPTION REFRIGERATION

Another type of refrigeration system which is much less common than the compressor type of refrigerator utilizes absorption. A liquid refrigerant is exposed to an absorbent which absorbs vapors from the refrigerant, causing it to evaporate. As it evaporates it absorbs the heat of vaporization. Continuous absorption refrigeration systems are much more complicated than compressor types, but operate on the same principle of utilizing the vaporization of a refrigerant to absorb heat.

REFRIGERATION OUTPUT

The output of a refrigeration system equals the amount of heat extracted from the cold reservoir. This is accomplished by doing work on the system. This leads to the term "coefficient of performance" or C.O.P.

C.O.P. = (heat extracted from refrigerated system) divided by (energy required to operate refrigerating machine)

The C.O.P. is a measure of the efficiency of the refrigeration system where a high C.O.P. means high efficiency. Theoretical C.O.P.s range from about 2.5 to more than 5.0.

If heat in the amount of Q_{ex} is extracted from a cold reservoir, and work in the amount of Q_w is done by the refrigerating machine--

C.O.P. = Q_{ex} / Q_w

The Carnot cycle is often used to describe a refrigeration machine. The Carnot cycle is described in the FE Handbook. When run in a reverse manner, a Carnot engine can act as a refrigerator. In this mode it extracts heat, Q, from a reservoir at T_2, does work, W, on it, and then discharges the heat and the work, Q + W, to a hotter reservoir at T_1. Since the refrigerant has gone through a complete cycle, there is no change in entropy.

$$C.O.P. = Q/W = T_2/(T_1 - T_2)$$

This is the Carnot C.O.P. The maximum C.O.P. which can be achieved between these two temperatures.

EXAMPLE 12-1

A Carnot refrigerator extracts heat from a cold-storage plant which is maintained at -30 C and discharges the heat to the outside which is at 38 C. If the refrigerator has a capacity of 100 U.S. tons of refrigeration, what power would be required to operate it?

Solution

The heat extracted is equal to $T_L \cdot \Delta S$ (see diagram for Carnot cycle in the FE Handbook). Work in the amount of--

$(T_H - T_L)\Delta S$ is required to operate the refrigerator machine. The amount of heat extracted from the cold source equals-- $(273 - 30) \times \Delta S = 243 \times \Delta S$ while only--

$311 - 243 = 68 \times \Delta S$ work is required to operate the engine. Thus the C.O.P. equals $243/68 = 3.57$ or 3.57 J of heat would be removed from the storage plant for every joule of work done by the engine. One hundred U.S. tons of refrigeration equals 351.7 kW so power of--

$351.7/3.57 = 98$ kW would be required.

EXAMPLE 12-2

In a refrigeration cycle a refrigerant flows through the compressor at the rate of 90 kg per hour. Calculate the C.O.P. if the temperature in the evaporator is -18 C and the temperature in the condenser is 38 C.

Solution

It will be assumed that compression of the refrigerant gas is isentropic, and that the gas out of the evaporator is dry and saturated. It is also assumed that there are no undetermined changes in properties of the refrigerant as it passes from one component to another in the refrigerant system. From data tables for this refrigerant--

For a temperature of 38 C the saturation pressure equals 910 kPa.

At -18 C $h_g = 179$ kJ/kg for dry saturated gas

Assuming a constant entropy compression from dry saturated gas at -18 C to 910 kPa of pressure gives the enthalpy for the superheated gas of 214 kJ/kg. This will cool to 38 C and condense to a liquid in the evaporator where the heat content of the fluid $h_f = 72$ kJ/kg

$Q_{in} = (h_g$ out @ -18 C) - (h_f into throttle valve at 38 C)

$= 179 - 72 = 107$ kJ/kg

For the compressor--

$W = h_g$ (superheat gas heat content) - h_g @ 18 C

$W = 214 - 179 = 35$ kJ/kg

C.O.P. = (heat into evaporator)/(work into compressor)

C.O.P. = 107/35 = 3.057

Check the max. Carnot C.O.P. between the two temperatures--

Max. C.O.P. = (273 - 18)/(311 - 255) = 4.55, so calculated C.O.P. is reasonable.

EXAMPLE 12-3

A mechanical refrigeration system utilizes ammonia as a refrigerant. Saturated vapor at -12 C flows to the compressor from the evaporator where it is compressed isentropically to 1.241 MPa. The compressed ammonia is liquified in the condenser at a temperature of 21 C. Assume no indeterminate losses in the system. Determine:

a) The amount of refrigeration.
b) The amount of heat rejected in the condenser.
c) The work by the compressor.
d) The C.O.P.

Solution

Data for ammonia:

Heat content for superheated gas compressed isentropically from -12 C to 1.241 MPa h_g = 1.670 MJ/g

Heat content of dry gas at -12 C h_g = 1.431 MJ/kg

Heat content of fluid in condenser at 21 C h_f = 281 kJ/kg

The refrigeration effect would equal--

h_g @ -12 C - h_f @ 21 C since there would be a constant heat expansion through the throttling valve

Refrigeration effect = 1,431 - 281 = 1,150 kJ/kg a)

The refrigerant enters the condenser as a superheated gas with h_g =1,670 kJ/kg and leaves as fluid with h_f = 281 kJ/kg

Heat rejected in condenser--

ΔQ = 1,670 - 281 = 1,389 kJ/kg b)

The compressor takes in gas with h_g = 1,431 kJ/kg and discharges superheated gas with h_g = 1,670 kJ/kg so it does work in the amount of 1,670 - 1,431 = 239 kJ/kg c)

The C.O.P. = cooling effect/work by compressor--

C.O.P. = 1,149/239 = 4.81 d)

This compares with the max. Carnot C.O.P. of $T_2/\Delta T$, Referring to the Carnot diagram in the FE Handbook--

max C.O.P. = $T_L/(T_H - T_L)$ = 261/33 = 7.91

HVAC

Heating, ventilating, and air conditioning is the science of creating a desired environment in a particular space. This may be a home, office, business area, work area, or storage area. The first three are primarily for personal comfort. Temperature control in a work area may also be necessary for dimensional stability, and control of the relative humidity (percentage humidity) in a storage area is very important in controlling rusting of steel or iron equipment.

BODY HEAT

The human body releases both sensible heat, since the ambient temperature is usually below the normal body temperature of 37.0 C, and latent heat due to perspiration. Perspiration adds water vapor to the air in the form of superheated steam which must be condensed to be removed. It is necessary to remove the superheated steam to maintain the humidity of the air-conditioned space at a desired value. The amount of heat required to accomplish the necessary condensation is termed the "latent heat gain" of the air-conditioned space. The rates of heat given off by an average body have been measured for various situations and conditions and are used to calculate cooling requirements for various circumstances.

EXAMPLE 12-4

A theater holds 600 people for an evening performance. Measurements have indicated that an average person's body gives off 205.7 kJ of sensible heat per hour (57.1 watts) and 163.5 kJ of latent heat per hour (45.4 watts), when sitting in as theater in the evening. What would be the refrigeration requirement for this theater to maintain a comfortable temperature during an evening performance?

SOLUTION

Six hundred people will give off heat at the rate of--

$600 \times (57.1 + 45.4) = 61{,}500$ watts or 61.5 kW

which is the amount of refrigeration required to counteract the heat supplied by the bodies in the theater.

SOLAR HEAT

Heat from the sun is an appreciable factor in the amount of air-conditioning required for a building. The latitude in which a building is located will have an effect on the amount of cooling required, so it is necessary to determine the latitude at which a building is located before calculating the heat load.

EXAMPLE 12-5

A building in Cleveland, Ohio is 30 m square and its walls are 3 m high. It is so oriented that the walls face north, east, west, and south. It has bare, single-pane windows in each wall with areas equal to ten percent of the wall area. What would be the maximum rate of solar heat gain through the east facing wall if the wall is painted a dark green on the outside? The wall is 20 cm thick brick with a plastered interior.

SOLUTION

From a map it is determined that Cleveland is at a latitude of 42 degrees. From a table for 40 degree latitude it is seen that the maximum rate of heat flow for a dark-colored wall facing east occurs at 10:00 A.M. standard time, and has a TD (solar temperature difference) of 6.7 C. The solid wall area equals $0.90 \times 90 = 81\ m^2$. The heat transfer coefficient for the wall has been determined to be:

$U = 1.703\ W/m^2{\cdot}K$

solar heat gain through wall--

$Q = 81 \times 1.703 \times 6.7 = 942$ watts over and above the heat gain by conduction.

Solar heat TD for a single-pane window at 40 degree latitude at 10:00 A.M. facing east equals 105 from an appropriate table, down from 173 at 8:00 A.M. The maximum excess solar heat gain through the window area would occur two hours earlier than for the solid wall area. The heat transfer coefficient for single-pane glass for solar heat gain equals $6.25\ W/m^2{\cdot}K$. It is interesting to note that the same heat transfer coefficient should be used for double-paneled windows since the solar heat gain through double-paned windows is almost the same as for single-paned windows.

$Q = A \times U \times TD$--

$9.0 \times 6.25 \times 173 = 9{,}731$ W at 8:00 A.M.

What would be the effect of changing the color of the outside wall to white and putting awnings over the windows?

The maximum TD for a white wall would drop from 44 to 13, a 70 percent reduction. For a window the maximum TD would drop from 173 to 52, again a 70 percent reduction.

VENTILATION

Air in an occupied room tends to get stale. To keep a space comfortable it is necessary to continuously replace a portion of the stale air with outdoor air. For auditoriums, theaters, etc. an air flow of 140 to 210 ℓ/min per person is recommended, for shops and lunch counters the recommended air flow rate rises to about 300 ℓ/m per person, and to some 700 ℓ/m per person for meeting rooms where there is heavy smoking.

PSYCHROMETRY

Psychrometry is the measurement of the relative humidity, RH, of the air RH equals the amount of water vapor present in the air divided by the greatest amount of water vapor the air could contain at the same temperature. The relative humidity of the air in a room very definitely affects the comfort of those occupying the space, and in HVAC operations it is necessary to control the RH as well as the temperature of the air.

EXAMPLE 12-6

An air-conditioned room is maintained at 25 C and 50% RH. The air-conditioning system recirculates 285 m^3/min. It cools 70% of the air and recycles 30%. The cooled portion leaves the cooling coils at 9 C dry bulb and 90% RH. The two flows are then combined and returned to the room. What are the resulting dry bulb and wet bulb temperatures of the mixture?

SOLUTION

The returned air dry bulb temperature equals--

$(0.70 \times 282 + 0.30 \times 298) \times 285/285 = 286.8$ K or 13.8 C The wet bulb temperature can be obtained from a psychrometric chart by drawing a line from the point at the intersection of the (50% humidity and the 25 C dry bulb temperature lines) and the point at the intersection of the (90% humidity and the 9 C dry bulb temperature lines). The point where this line intersects with the dry bulb temperature line of 13.8 C show a wet bulb temperature of 12.1 C for the mixture of the two air flows.

EXAMPLE 12-7

A line carrying dry saturated steam at 790 kPa pressure passes through an instrument room 6m × 6m × 4m. The steam line develops a leak and steam leaks into the room at the rate of 1.0 kg per hour and diffuses thoroughly throughout the space. The room is initially at a temperature of 20 C and has a RH of 50%. The air pressure is one standard atmosphere. Estimate what the RH would be after one hour, assuming no exchange of heat between the room and its surroundings? The vapor pressure of saturated steam at 20 C is, 2.337 kPa.

Temperature of saturated steam at (790 + 101) kPa = 175 C

SOLUTION

Assume the steam and the air are perfect gasses.

The volume of the room equals $6 \times 6 \times 4 = 144\ m^3$

Initially the water vapor in the room will exist as a superheated gas.

The steam would expand through the hole in the line to atmospheric pressure. The expansion would be constant heat since the velocity of the jet would soon be dissipated and the kinetic energy re-converted. The heat added to the room air would equal--

$1.0\text{ kg} \times 1.87\text{ kJ/(kg·K)} \times (175 - 20)\text{ K} = 290\text{ J}$

The molecular weight of steam is 18. The universal gas constant equals 8,314 J/(kmol·K) (from the FE handbook) so the gas constant for steam equals--

$R_{Steam} = 8{,}314/18 = 463.4$ J/K

If the room were filled with saturated steam at 20 C, RH of 100%, the partial pressure of the steam would equal 2,337 Pa, but the RH is only 50%, so the partial pressure of the steam is 1,169 Pa. Another way of looking at this is to assume that half of the volume is filled with saturated steam at 2,337 Pa. Either approach is permissible and satisfies the requirements of partial pressures and both will give the same answer.

PV = mRT or m = PV/RT so--

m_{steam} = 1,169 × 144/(463.4 × 293) = 1.240 kg of superheated steam at initial conditions.

R_{air} = 8,314/29 = 286.69

m_{air} = (101,300 - 2,337) × 144/(286.69 × 293) = 169.7 kg of air at initial conditions.

The addition of 290 joules of heat would increase the temperature of the air mixture in the room by 290/(169.7 × 1.00) = 1.71 C which is small compared with the temperature of 293 C, so that its effect on the humidity in the room can be neglected for a first approximation.

Relative humidity equals the mass of water vapor held in the air divided by the maximum mass of water vapor which could be held in the same volume of air without any of the water vapor's condensing. The RH in this case is then proportional to the mass of water vapor held in the air in the room. The mass of water vapor would be increased by 1.00 kg after one hour, so the RH after one hour would equal

RH = (1.240 + 1.00)/1.240 times 50% or 90%

This is a reasonably accurate estimate and an estimate is what has been requested.

EXAMPLE 12-8

A research laboratory building is to be heated and ventilated. Air is to be supplied to a room 12 m by 18 m from the outside at the rate of one m³/min air replacement for each square meter of floor space. The design conditions include 20 C room temperature and -20 C outside air temperature. In addition, the pressure in the room is to be maintained at a positive pressure of 7.0 kPa, to ensure that there will be no air leakage into the laboratory. How much heat would be required to condition the air for this room?

SOLUTION

The floor area equals 216 m² so an air flow rate of 216 m³/min would be required at (101,300 + 7,000) Pa absolute pressure. The mass of air--

m = PV/RT = 108,300 × 216/(286.7 × 293) = 278 kg

(where R = 8,314/29 = 286.7)

Q = 278 kg/min × 1.00 kJ/(kg·K) × 40 K = 11,120 kJ/min

11,120/60 = 185.3 kJ/s or 185.3 kW heating required.

PROBLEMS

12-1. An office air-conditioning system drew in a quantity of hot air from outside, cooled it to 1 C, and mixed it with the internal air of a building. Care was taken to supply enough fresh air to eliminate objectionable odors. The temperature of the space was maintained constant, and yet some people became uncomfortable in the afternoon. Why?

a) The relative humidity had dropped.
b) When a person spends a long time at a constant temperature that person becomes uncomfortable.
c) Radiant heat from others in the office made the space uncomfortable.
d) The complainers were not really uncomfortable, they were just bored.

12-2.
What would be the cooling requirement due to the emission of body heat to maintain a comfortable temperature in a dance hall that had an attendance of 1,500 people if it has been determined that 75% of the people will be dancing and the others will be seated. Tests have shown that an adult will give off 71.8 W of sensible heat while engaging in moderate dancing, and 57.1 W of sensible heat while sitting. The latent heat emissions for similar situations have been found to equal 177.3 W for dancers, and 45.4 for sitters.

a) 200 kW
b) 267 kW
c) 301 kW
d) 319 kW

12-3. A refrigerator is rated at a U.S.A. one ton capacity. It maintains a cooling temperature of 4.4 C (evaporator temperature). A refrigerant with a refrigerating effect of 136 kJ/kg is in the condenser just ahead of the expansion valve. If the refrigerant vapor has a density of 19.80 kg/m³ what should be the capacity of the compressor?

a) 0.078 m³/min
b) 0.091 m³/min
c) 0.102 m³/min
d) 0.113 m³/min

SOLUTIONS

12-1. Spending time at a constant temperature, other things being equal, does not, in itself, make a person uncomfortable. Heat radiates from a hotter body to a cooler one, and even if there were feverish people in the office the amount of heat absorbed by radiation would be negligible. The complainers could have been bored, but if moisture were not added to the office air it is quite possible that the dryness of the air could cause some discomfort. When the outside air was cooled to 1 C, it became saturated and water contained in the air condensed. Then, when the air was heated, the relative humidity of the heated air was lower than that of the air in the office. Over a period of a few hours this could reduce the relative humidity to the point where it was no longer comfortable. For comfort, at a temperature between 21 and 24 degrees C, the relative humidity should be in the range of 30% to 60% for most people. A well designed complete air-conditioning system incorporates equipment to maintain the relative humidity in the comfort range. The most probable cause of the discomfort was a reduction in the relative humidity.
The correct answer is a).

12-2.
The total bodily emitted heat would thus equal--

$$1{,}125 \times (71.8 + 177.3) + 375 \times (57.1 + 45.4) =$$
$$280{,}328 + 38{,}438 = 318{,}676 \text{ W or } 319 \text{ kW}$$

The correct answer is d).

12-3. One ton of refrigeration is the amount to freeze one U.S. ton of water (907 kg) in 24 hours--
907 kg × 335 kJ/kg × 1/(24 × 3,600) = 3.52 kW
3.52/136 = 0.0259 kg/sec

$$Q = (0.0259/19.80) \times 60 = 0.0784 \text{ m}^3/\text{min}$$

The correct answer is a).

Chapter 13

Fans, Pumps, and Compressors

FANS

A fan is essentially an air pump. It moves a gas, usually air. Gasses are compressible, but at low pressures can be considered as incompressible for engineering calculations. At low velocities the effect of compressibility is negligible and is approximately equal to one-quarter the square of the Mach Number. So for velocities less than 100 m/s air can be considered as being incompressible with negligible error.

FAN CHARACTERISTICS

There are a number of characteristics of fan operation that can affect their application. Generally speaking the following the following are true:

1) For a given fan the volumetric output will vary directly with the speed of rotation.

2) The head, or pressure, supplied by the fan will vary as the square of the speed of the fan.

3) For a given fan, the power required to operate the fan will vary as the cube of the speed.

4) For a given installation and constant fan speed, the pressure output and the operating power requirement will vary as the density of the gas. That is, the head provided by the fan will remain constant so the pressure and power will be proportional to the density of the gas. The volumetric output will remain constant for the same speed.

5) For a constant mass flow rate of gas, the speed of the fan and the volumetric output will vary inversely as the density of the gas. For a constant fan speed the power required will vary inversely as the square of the density.

EXAMPLE 13-1

A boiler that burns coal requires 1,700 m³ of air per minute at a pressure of 15 cm of water gage for the combustion process. If the air is supplied by a fan which has a mechanical efficiency of 58% at these conditions, what size motor would be required to operate the fan?

Solution:

For this case the air can be considered as an incompressible fluid, since the effects of compressibility will be negligible at this low pressure.

The fan power output will equal volume per second times pressure which equals newton-meters/s, or watts.

The flow rate would equal 1,700/60 = 28.33 m³/s

The pressure differential of 15 cm of water

0.15 m height of one cubic meter of water which has a mass of 1,000 kg giving a mass of 150 kg which equals a force of 1,471 N over an area of one square meter, giving a pressure of 1,471 Pa

Air power = 28.33 × 1,472 = 41.70 kW

Motor power = 41.70/0.58 = 71.90 kW

EXAMPLE 13-2

If the requirement for air for combustion in the preceding problem were given for standard conditions of one atmosphere and 20 C, but the local temperature was 30 C how would this a) affect the volumetric requirement, and b) the power requirement?

Solution:

The amount of air for combustion of a quantity of coal is given as a volume in the example above, but the actual requirement for combustion is the mass of oxygen. If the air used for the combustion does not contain the required mass of oxygen, the volume of air must be increased to provide the mass of oxygen necessary. The gas law states--

PV = MRT for this case M, the mass of air supplied, is to be constant as are the gas constant R, and the pressure P, so--

$V_2 = V_1 \times (T_2/T_1) = 1{,}700 \times 303/293 = 1758\ m^3/min$ a)

The power required by the fan is proportional to the flow rate for the same pressure output so the power required would increase in the same ratio--

$P_2 = 71.90 \times (303/293) = 74.35\ kW$ b)

The power can also be calculated by multiplying the mass rate of flow by the meters of head supplied by the fan.

$M = PV/RT = (101.3\ kPa \times 29.3\ m^3/s)/(286.7 \times 303)$

$M = 1.166\ kg/m^3 \times 29.3\ m^3/s = 34.164\ kg/s$

Pressure rise = 1,472 Pa or 1,472 N/m^2

specific mass $N/m^3 = 1.166 \times 9{,}8066\ N/kg = 11.435\ N/m^3$

ΔH = head supplied by fan = 1,472/11.435 = 128.7 m

Air power = $34.164 \times 9.8066 \times 128.7 = 43.12$ kW

Power required = 43.12/0.58 = 74.34 kW as before

EXAMPLE 13-3

A ventilating operation has a fan which delivers 400 m^3/min against a back pressure of 5 cm water gage when the fan runs at 450 rpm. A new requirement calls for an increase in air flow to 500 m^3/min. What is the present air power produced by the fan, a), what will be the new speed, b), the new back pressure, c), and the new air power?

Solution:

The flow rate equals 400/60 = 6.667 m^3/s

The back pressure equals--

$0.05\ m \times 1{,}000\ kg \times 9.8066 = 490.33\ N/m^2$ or 490.33 Pa

Air power = $6.667 \times 490.33 = 3{,}269$ W or 3.27 kW a)

The new speed required will equal--

$450 \times 500/400 = 562.5$ b)

The new back pressure will equal--

$5.0 \times (500/400)^2 = 7.97$ cm water gage c)

The new air power will equal--

$3.27 \times (500/400)^3 = 6.39$ Kw

EXAMPLE 13-4

A blower delivers 350 m^3/min at standard barometric pressure and 20 C when rotating at 500 rpm against a back pressure of 5 cm of water gage. If the air temperature should rise to 90 C, and the speed of the fan remained constant, what would the back pressure become, a) and what would be the ratio of the power required to the original power requirement, b)?

Solution:

The speed of the fan would stay the same so the volume flow would be the same, since from the list of fan characteristics volumetric output varies directly with the speed of rotation. The head supplied by the fan, however, will vary as the density of the gas. The density at 90 C, ρ, will equal P/RT, or the density of a gas (air) is inversely proportional to the absolute temperature, other factors remaining the same. The density at the higher temperature would equal 293/363 = 0.807 times the original density so the new back pressure would equal--

0.807 × 5.0 = 4.04 cm of water gage a)

At constant speed and volumetric output the power varies directly as the density of the gas, so the power required would equal 80.7% Of the original power requirement. b)

The original air power--

5.833 m^3/min × 490.33 Pa = 2.86 kW

EXAMPLE 13-5

If the speed of the fan in the previous example is increased so that the back pressure is maintained at 5 cm of water, what would be the new speed, a), what would be the new air flow rate, b), and how would the power requirement be affected. c)?

Solution:

For a constant pressure differential, constant back pressure in this case, the speed, flow rate, and power, will all vary inversely as the square root of the density of the gas.

The ratio of the densities as calculated above equals 0.807. Then the new speed requirement equals--

500 × √(1/0.807) = 500 × 1.113 = 557 rpm a)

Similarly the new flow rate--

Q = 350 × 1.113 = 390 m^3/min (of 90 C air) b)

The new fan power requirement--

Power = 1.113 × 2.86 = 3.18 kW

EXAMPLE 13-6

If it is desired to maintain the same mass flow rate of air at 90 C as at 20 C with the above fan installation, what would be the required speed, a)? What would be the new flow rate, b), the back pressure, c), and the new power, d)?

Solution:

From the list of fan characteristics, to maintain a constant mass flow rate for air requires that the speed of the fan vary inversely as the density of the air at the two conditions. The volumetric flow and the back pressure will both vary inversely as the ratio of the densities. Power is proportional to flow rate times back pressure so the new power requirement will equal the square of the inverse ratio of the densities. The ratio of the densities at the two conditions has been calculated above and equals 0.807 .

New speed = 500/0.807 = 620 rpm a)

New air flow rate = 350/0.807 = 434 m^3/min b)

New back pressure = 5/0.807 = 6.2 cm water gage c)

New power requirement--

2.86 × $(1/0.807)^2$ = 4.39 kW fan power d)

PUMPS

A pump is a device that is used to increase or decrease the pressure of a fluid. A fan is a special type of pump for pumping air. A similar type of pump for pumping a liquid is an axial-flow pump.

There are many types of pumps, but the two commonest types for engineering purposes are centrifugal pumps and positive-displacement pumps.

A centrifugal pump consists of a circular impeller which rotates within a case. The liquid enters at the center portion of the impeller and is accelerated outward due to centrifugal force. The fluid thus absorbs kinetic energy from the impeller, part of which is converted to pressure as it reaches the outer periphery of the impeller and is discharged out of the case. The efficiency of a centrifugal pump equals--
(fluid power out of pump)/(power used to operate the pump) or the efficiency of converting the power applied to the shaft divided by the power contained by the fluid as it leaves the pump.

EXAMPLE 13-7

A pump with an inlet pipe 25 cm in diameter and a discharge pipe 15 cm in diameter absorbed 12 kW of power at 1500 rpm. The pump output was 3,100 liter/min of 20 C water at 110 kPa. The pressure measured at the suction side of the pump was a minus 12 cm of mercury, and the center line of the discharge pipe was 1.0 m above the center line of the intake pipe. What was the efficiency of the pump?

Solution:

Bernoulli's equation can be applied to obtain the energy added to the liquid.

$$P_1/\gamma_1 + Z_1 + V_1^2/2g + E = P_2/\gamma_2 + Z_2 + V_2^2/2g + F$$

where E = energy added and F = energy lost due to friction
Calculating the individual terms gives--

(friction loss through pump assumed to equal zero)

$P_1 = -(12/76) \times 101.3\ \text{kPa} = -15.995\ \text{kPa}$

$P_1/\gamma = -15{,}995/(1{,}000 \times 9.8066) = -1.631\ \text{m}$

$P_2/\gamma = 110{,}000/(1{,}000 \times 9.8066) = 11.217\ \text{m}$

$Z_2 - Z_1 = 1.00\ \text{m}$

$A_1 = 0.0491\ \text{m}^2 \quad v_1 = (3.100/60)/0.0491 = 1.0523\ \text{m/s}$

$V_1^2/2g = 1.1073(2 \times 9.8066) = 0.0565\ \text{m}$

$A_2 = 0.0177\ \text{m}^2 \quad v_2 = 2.919\ \text{m/s}$

$v_2^2/2g = 8.5207/2g = 0.4344\ \text{m}$

Combining terms gives--

$-1.631 + 0.0565 + \text{energy added} = 11.217 + 1.00 + 0.4344$

Energy added = 14.226 m

Pump power--

$[(3{,}100/60) \times 9.8066]\ \text{n/s} \times 14.226\ \text{m} = 7{,}208\ \text{J/s}$

Pump power = 7.208 kW

Efficiency = 7.208/12 = 0.601 or 60.1 percent

AFFINITY LAWS

For a centrifugal pump the following laws apply:

1) Flow rate is proportional to speed of impeller.
2) Head produced is proportional to speed squared.
3) Power is proportional to speed cubed.

In addition, for a particular pump for small changes in impeller diameter:

1) Capacity varies directly as impeller diameter.
2) Head varies as the square of the impeller diameter.
3) Power varies as the cube of the impeller diameter.

The ratio of power follows directly from those of flow rate and head since power equals flow rate times head.

EXAMPLE 13-8

If the 1,500 rpm motor in the preceding example were replaced by a 1,750 rpm motor, what would the new flow rate be, a), what would be the new output pressure, b), and what would be the new power requirement, c)? Assume the efficiency of the pump at the higher speed would be the same as at the lower speed.

Solution:

New flow rate = 3,100 × (1,750/1,500) = 3,617 ℓ/min a)
New output pressure = 110 × 1.1667^2 = 150 kPa
New power requirement = 12 × 1.1667^3 = 19 kW

CAVITATION

When the local pressure of a liquid falls below its vapor pressure the liquid will vaporize and a cavity or void will form at that point. The implosion of such cavities can be quite violent, and often causes pitting in impellers and ship propellers. Cavitation is usually accompanied by noise resulting from collapse of the bubbles and creates havoc with the efficiency of a pump. As a result, care should be taken that the pressure in the suction line of a pump be maintained above the vapor pressure of the liquid being pumped at the temperature in the suction line.

EXAMPLE 13-9

Water at 65 C is drawn up an intake line to a pump inlet with an average velocity of three m/s. At one point in the vertical line the pressure equals 7.0 kPa. At what distance above this would cavitation start?

Solution:

At a temperature of 65 C the vapor pressure of water is 188 mm of Hg, and the density is 0.981 g/mℓ or 981 kg/m³.
188/760 × 101.3 kPa = 25.06 kPa vapor pressure of water
The absolute pressure at the reference point in the intake line P_a = 101.3 + 7.0 = 108.3 kPa. This pressure would have to drop 108.3 - 25.1 = 83.2 kPa for the pressure to drop to the vapor pressure, at which point cavitation would take place. 83,200/(981·9.8066) = 8.65 m. This is the maximum height above the reference point that the pump intake could be located to prevent cavitation.

N.B The total head in the line would equal the measured pressure plus the velocity head, but since the fluid velocity would be the same at the two points, this factor cancels out. The data on density and vapor pressure of water are not available in the FE Handbook. These data would be provided to the examinee as necessary.

NPSH

Cavitation can also occur in the impeller of a centrifugal pump. To prevent its happening a pump manufacturer tests his pump design to determine just how much pressure is necessary at the intake of the pump to prevent cavitation during pump operation. This is termed the "Net Positive Suction Head" which is abbreviated NPSH.

$$NPSH = P_a/\gamma + v^2/2g - h_v$$

Where: P_a = absolute pressure at pump intake
$v^2/2g$ = velocity head in the suction line

h_v = absolute vapor pressure head of the liquid at the operating temperature

EXAMPLE 13-10

Liquid propane is contained in a storage tank at a pressure of 1.4 MPa_a, which is the equilibrium vapor pressure of the propane at the pumping conditions. The propane has a specific gravity of 0.58. If the level of the liquid propane is 4.00 m above the center line of the pump and the frictional losses in the suction line equal 2.00 m, what is the available NPSH if the velocity in the suction line is 1.50 m/s?

Solution:

The net pressure at the pump intake would equal--

P = 1.4 MPa_a + 4.00 m - 2.00 m - 1.4 MPa = 2.00 m plus a velocity head of $1.5^2/(2 \cdot 9.8066)$
= 0.115 m of propane

The available NPSH = 2.12 m.

EXAMPLE 13-11

A pump discharges 7,500 liters/min of a brine with a specific gravity of 1.20 to an evaporation pond. The intake line is 30.00 cm in diameter, and is at the same level as the pump discharge line which is 20.00 cm diameter. The pressure at the pump inlet is minus 150 mm of mercury. The pressure gage connected to the pump discharge reads 140 kPa and its center is 1.50 m above the center of the discharge flange. If the pump efficiency is 82 percent, what power is required to operate it? If the vapor pressure of the brine is 35.3 mm Hg, what is the NPSH? If the pump output is increased to 10,000 ℓ/min, what would be the new NPSH?

Solution:

Use Bernoulli's equation to solve for the energy added to the brine.

$$P_1/\gamma + Z_1 + V_1^2/2g + E = P_2/\gamma + Z_2 + V_2^2/2g + F$$

where E = energy added and F = energy lost due to friction.

P_1 = -150 mm Hg = -(150/760) × 101.3 kPa = -19.99 kPa

The density of the brine equals 1.2 × 1,000 = 1,200 kg/m³

$P_1/\gamma = -19{,}990/(1{,}200 \cdot 8.9066) = -1.699$ m

$Z_1 = Z_2$

$A_1 = 0.0707$ m² $v_1 = (7.50/60)/0.0707 = 1.7680$ m/s

$v_1^2/2g = 0.159$ m

$P_2/\lambda = (140{,}000/(1{,}200 \cdot 9.8066) + 1.5 = 13.397$

$A_2 = 0.0314$ m² $v_2 = 3.981$ m/s $v_2^2/2g = 0.8080$ m

E = 13.397 + 0.808 + 1.699 - 0.159 = 15.745 m added

(7.500/60) × 1,200 × 9.8066 × 15.745 = 23,161 W or 23.161 kW

Power required to drive pump = 23.161/0.82 = 28.245 kW

The NPSH equals--

$P_a/\gamma = [101{,}300/(1{,}200 \cdot 9.8066)] - 1.699$
= 8.6081 - 1.699 = 6.9091 m

$v^2/2g = 0.159$ $h_v = (35.3/760) \times 8.6081 = 0.400$

NPSH = 6.909 + 0.159 - 0.400 = 6.668 m

If the flow rate were increased to 10,000 ℓ/m the pressure drop in the suction line would increase in the ratio--

$(10{,}000/7{,}500)^2 = 1.778$ as would the velocity head.

For the new NPSH--

The negative intake absolute pressure, meters of brine

$P_a/\lambda = 8.6081 - (1.778 \times 1.699) = 5.587$ m
$v^2/2g = 1.778 \times 0.159 = 0.283$ m
The new NPSH = 5.587 + 0.283 - 0.400 = 5.470 m

COMPRESSORS

Compressors are usually positive displacement pumps with reciprocating pistons. The capacity of an air compressor is usually stated as the quantity of air handled at intake conditions.

EXAMPLE 13-12

For a compressor rated at 20.0 m³/min, at standard conditions, how much air compressed air at 1.70 MPa and 25 C would be delivered by the compressor?

Solution:

Take standard conditions as 20 C and one atmosphere.
The perfect gas law states PV = mRT. Since the gas constant and the mass of air are constant for this case

$P_1 \cdot V_1/T_1 = P_2 \cdot V_2/T_2$ giving--
$V_2 = 101.3/(1{,}700 + 101.3) \times 298/293 \times 20.0 = 1.143$ m³/min

EXAMPLE 13-13

Assuming isothermal compression and a compressor with no clearance, how much work would be required to compress 150 m³/hr of "standard air" to 700 kPa in a single-stage compressor?

Solution:

This would be an example of flow work since air would be "flowing" through the compressor-- $W = \int V \cdot dP$
for isothermal conditions $P \cdot V$ = constant, or $V = mRT/P$
and $W = mRT \ln(P_1/P_2)$
or, since $m = PV/RT$ $W = P_1 \cdot v_1 \times \ln(P_1/P_2)$

$W = 101{,}300 \times 150 \times \ln(101.3/801.3) = -31.43 \times 10^6$ J, or work done on the air.

PROBLEMS

13-1. The volumetric output of a centrifugal pump is reduced by closing a valve part way on the intake side of the pump, throttling the intake. The reduction of flow is accompanied by "popping" noises which seem to come from inside the case. What is the probable cause?

a) The impeller has broken.
b) Dirt particles have got into the pump intake.
c) The intake valve flow seal is loose and vibrates.
d) The fluid is cavitating.

13-2. A centrifugal pump pumps water at the rate of 1,000 ℓ/min against a head of ten m. What is the power requirement?

a) 1.59 kW
b) 1.63 kW
c) 1.98 kW
d) 1.07 kW

13-3. A fan delivers 225 m^3 of standard air per minute against a back pressure of 10.0 cm of water to a coal burning furnace. What power is required to operate the fan if it is 85% efficient?

a) 4.33 kW
b) 3.68 kW
c) 3.95 kW
d) 4.87 kW

SOLUTIONS

13-1. The most probable cause is that the fluid being pumped is cavitating since the negative pressure on the intake side of the pump has been increased. Throttling the intake will reduce the output of a pump, but it lowers the pressure on the fluid being pumped and can reduce it below the NPSH requirement for the particular pump at the intake, thus inducing cavitation accompanied by the noise from the collapsing vapor bubbles. It is best to control pump output by throttling the discharge.
The correct answer is d).

13-2. The pump pumps 1,000/60 = 16.667 liters/sec wit a mass of 16.667 kg against a head of 10.0 m. The energy requirement equals--

$$16.667 \times 9.8066 \times 10.0 = 1{,}634 \text{ W} \quad \text{or} \quad 1.634 \text{ kW}$$

The correct answer is b).

13-3. Standard air has a density of--
$m = PV/RT = 101.3 \text{ kPa} \times 1.0/[(8.314/29) \times 293] = 1.206 \text{ kg/m}^3$
$10.0 \text{ cm} \times 1{,}000/1.206 = 8{,}292 \text{ cm}$ or 82.92 m
mass flow $= 225 \times 1.206/60 = 4.523$ kg/sec
$4.523 \times 9.8066 \times 82.93 \text{ m} = 3{,}678 \text{ W}$ or 3.68 kW air power
$3.678/0.85 = 4.33$ kW required to operate the fan.
The correct answer is a).

AFTERNOON SAMPLE EXAMINATION

Mechanical	No. Probs.
Mechanical Design	6
Dynamic Systems,vibration,kinematics	6
Thermodynamics	6
Heat Transfer	6
Fluid Mechanics	6
Stress Analysis	6
Measurement and Instrumentation	6
Material Behavior/ Processing	3
Computer, Automation, Robotics and Numerical methods	3
Energy Conversion and Power Plants	3
Automatic control	3
Refrigeration and HVAC	3
Fans, Pumps, and Compressors	3

FUNDAMENTALS OF ENGINEERING EXAM

AFTERNOON SESSION

ⒶⒷⒸ Fill in the circle that matches your exam booklet

1 ⒶⒷⒸⒹ
2 ⒶⒷⒸⒹ
3 ⒶⒷⒸⒹ
4 ⒶⒷⒸⒹ
5 ⒶⒷⒸⒹ
6 ⒶⒷⒸⒹ
7 ⒶⒷⒸⒹ
8 ⒶⒷⒸⒹ
9 ⒶⒷⒸⒹ
10 ⒶⒷⒸⒹ
11 ⒶⒷⒸⒹ
12 ⒶⒷⒸⒹ
13 ⒶⒷⒸⒹ
14 ⒶⒷⒸⒹ
15 ⒶⒷⒸⒹ
16 ⒶⒷⒸⒹ
17 ⒶⒷⒸⒹ
18 ⒶⒷⒸⒹ
19 ⒶⒷⒸⒹ
20 ⒶⒷⒸⒹ
21 ⒶⒷⒸⒹ
22 ⒶⒷⒸⒹ
23 ⒶⒷⒸⒹ
24 ⒶⒷⒸⒹ
25 ⒶⒷⒸⒹ
26 ⒶⒷⒸⒹ
27 ⒶⒷⒸⒹ
28 ⒶⒷⒸⒹ
29 ⒶⒷⒸⒹ
30 ⒶⒷⒸⒹ
31 ⒶⒷⒸⒹ
32 ⒶⒷⒸⒹ
33 ⒶⒷⒸⒹ
34 ⒶⒷⒸⒹ
35 ⒶⒷⒸⒹ
36 ⒶⒷⒸⒹ
37 ⒶⒷⒸⒹ
38 ⒶⒷⒸⒹ
39 ⒶⒷⒸⒹ
40 ⒶⒷⒸⒹ
41 ⒶⒷⒸⒹ
42 ⒶⒷⒸⒹ
43 ⒶⒷⒸⒹ
44 ⒶⒷⒸⒹ
45 ⒶⒷⒸⒹ
46 ⒶⒷⒸⒹ
47 ⒶⒷⒸⒹ
48 ⒶⒷⒸⒹ
49 ⒶⒷⒸⒹ
50 ⒶⒷⒸⒹ
51 ⒶⒷⒸⒹ
52 ⒶⒷⒸⒹ
53 ⒶⒷⒸⒹ
54 ⒶⒷⒸⒹ
55 ⒶⒷⒸⒹ
56 ⒶⒷⒸⒹ
57 ⒶⒷⒸⒹ
58 ⒶⒷⒸⒹ
59 ⒶⒷⒸⒹ
60 ⒶⒷⒸⒹ

DO NOT WRITE IN BLANK AREAS

Instructions for Afternoon Session

1. You have four hours to work on the afternoon session. You may use the *Fundamentals of Engineering Reference Handbook* as your *only* reference. Do not write in this handbook.

2. Answer every question. There is no penalty for guessing.

3. Work rapidly and use your time effectively. If you do not know the correct answer, skip it and return to it later.

4. Some problems are presented in both metric and English units. Solve either problem.

5. Mark your answer sheet carefully. Fill in the answer space completely. No marks on the workbook will be evaluated. Multiple answers receive no credit. If you make a mistake, erase completely.

Work 60 afternoon problems in four hours.

1.
A coupling consisting of two circular flat plates welded to the ends of two shafts is to be designed using 2.54 cm diameter bolts to connect one plate to the other. The bolts are to be located at a distance of 0.1524 m from the axes of the two shafts. If the coupling is to transmit 4,026.8 kW at a shaft speed of 1,200 rpm, how many bolts should be used in the coupling if the allowable shear stress for the bolts is 103.43 Mpa?
A) two
B) three
C) four
D) five

2.
A hollow steel shaft 2.54 m long is to be designed to transmit a torque of 33.900 kJ. The shear modulus of elasticity, G = 82.740 GPa for steel. The total angle of twist in the length of the shaft must not exceed 3o and the stress in the shaft must not exceed 110.32 MPa. What would be the ratio of outside diameter (d) to inside diameter (d)?
A) 1.92
B) 1.74
C) 1.56
D) 1.38

3.
Four bolts are used to hold a bracket to a base as shown. The maximum shear load which can be applied to a bolt is 8,896 N. Based on bolt strength, what is the maximum load which can be applied to the bracket at the point "F"?
A) 8,896 N
B) 9,249 N
C) 9,602 N
D) 9,955 N

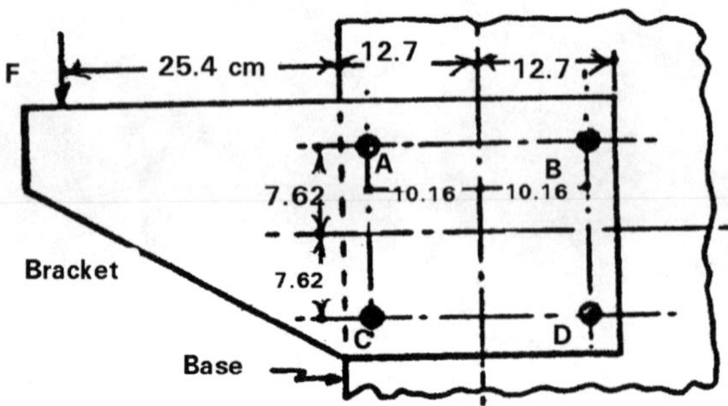

4.
A 19.05 mm diameter rod is used to hold up a wooden beam. The pitch of the rod thread is 2.54 mm. The rod passes through the beam and support is effected by a nut on the lower end. The root diameter of the thread is 15.75 mm. The tensile strength of the steel is 137.9 MPa. The compressive strength of the wood is 2.379 MPa. What diameter of washer should be placed between the nut and the wooden beam so that the wood is not over-stressed when the rod is carrying its maximum allowable load?
A) 97.3 mm
B) 110.5 mm
C) 123.7 mm
D) 136.9 mm

5.
A brake shoe is pressed against the surface of a 25.4 cm diameter drum rotating at 1,250 rpm with a force of 111.20 N. If the coefficient of friction between drum and cylinder is 0.21, what power is dissipated in the form of heat by the drag of the brake shoe?
A) 388.2 W
B) 394.9 W
C) 401.6 W
D) 408.8 W

6.
A freight car with a total total mass of 68,027 kg breaks loose at the top of a 304.8 m long 5° grade. The rolling resistance friction factor equals 0.050. It hits a bumper formed of a combination of heavy duty springs at the bottom of the grade. What must be the spring constant of the bumper assembly to bring the freight car to a stop in 1.0668 m? (Assume bumper track is horizontal.)
A) 13.34 MN/m
B) 13.12 MN/m
C) 12.90 MN/m
D) 12.69 MN/m

7.
A flat circular disk 0.500 m in diameter is made of steel (sp. gr. 7.87) and is 20 mm thick. It is attached solidly at its center to a 10.0 mm diameter steel rod 1.0 m long and is oriented horizontally. The upper end of the attached rod is rigidly held by a massive support. What is the period of vibration of this torsional pendulum?
A) 0.52 sec
B) 0.68 sec
C) 0.84 sec
D) 1.00 sec

8.
What is the tension in the rope for the system shown ? (Assume no friction and weightless Pulleys.)
A) 1.03 kN
B) 1.23 kN
C) 1.43 kN
D) 1.63 kN

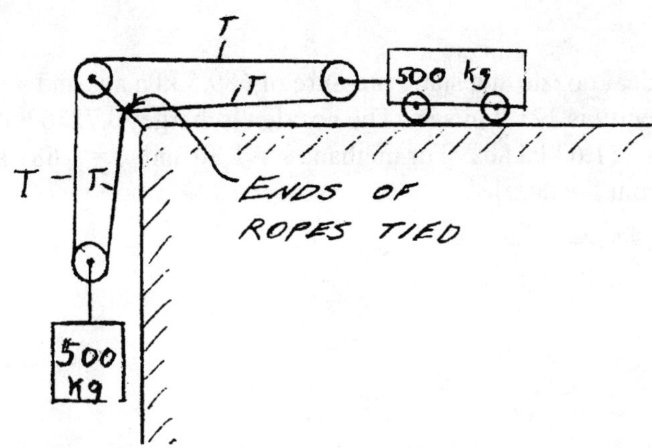

Figure for Problem 8.

9.
Air at 103,425 Pa absolute with a volume of 0.2831 m^3 is isentropically compressed by transfer of 63,300 J until the temperature is 190.6 C. Assume k = 1.4
R_{air} = 286.82 J/kg-K. What is the change in internal energy?
A) 69,848 J
B) 63,200 J
C) 57,196 J
D) 51,805 J

General: The necessary constants are listed in the FE Reference Handbook issued by the National Council of Examiners for Engineering and Surveying.

10.
A vessel with a volume of 1.416 m^3 is being filled with air, which is to be considered a perfect gas. At time zero the temperature in the tank is 115.56 C0 and the pressure is 1.379 MPa abs0. The pressure is increasing at a rate of 137.9 kPa/sec and the temperature is increasing at the rate 27.78 C/sec. What is the instantaneous rate of air flow into the tank at this instant of time?
A) 0.36 kg/sec
B) 0.41 kg/sec
C) 0.45 kg/sec
D) 0.50 kg/sec

11.
A fluid at 689.5 kPa and a specific volume of 0.250 m^3/kg enters an apparatus with a velocity of 152.4 m/sec. Heat and radiation losses in the apparatus equal 23,263 J/kg of fluid supplied. The fluid leaves the apparatus at a pressure of 137.90 kPa with a specific volume of 0.9365 m^3/kg) and a velocity of 304.79 m/sec. In the apparatus the shaft work done by the fluid is equal to 582,938 J/kg. What is the change in the internal energy of the fluid?
A) -598 kJ/kg
B) -561 kJ/kg
C) -523 kJ/kg
D) -486 kJ/kg

12.
Methane (CH_4) enters an ideal nozzle at a static pressure of 689.5 kPa abs and a static temperature of 26.67 C. The entering velocity is 121.9 m/sec. The nozzle discharges 0.7256 kg/sec into a vessel in which the static pressure is 551.6 kPa abs. For methane k = 1.30 and c_v = 1.687 kJ/kg○K.
What is the exit velocity from the nozzle?
A) 207 m/sec
B) 230 m/sec
C) 256 m/sec
D) 284 m/sec

13.
A heavy fuel oil with a specific gravity of 0.94 and a viscosity of 2.01 poise is to be pumped at the rate of 2,000 liters/min through a 30 cm I.D. pipe. What would be the friction factor?
A) 0.033
B) 0.055
C) 0.077
D) 0.099

14.
An ore carrier 120 m long by 1 m wide displaces 8,500 m^3 of fresh water. It is moved into a lock in fresh water that is 140 m long by 15 m wide and then is loaded with 3,500 metric tons of ore. What will be the increase in the depth of the water in the lock after the ship is loaded?
A) 0.83 m
B) 1.67 m
C) 2.33 m
D) 2.67 m

15.
A careless hunter shoots a 7.00 mm diameter hole in the vertical side of a water tank one meter above the ground. The level of the water in the tank is 10.0 m above the ground. How far from the base of the tank will the water strike the ground?
A) 2.67 m
B) 4.75 m
C) 6.00 m
D) 7.88 m

16.
The water level of a reservoir is 150 m above a power plant. A 30.0 cm I.D. pipe 200 m long connects the reservoir to the plant. If the friction factor for flow through the pipe is 0.02, how much water will flow to the plant?
A) 0.80 m^3/s
B) 1.0 m^3/s
C) 1.2 m^3/s
D) 1.4 m^3/s

17.
A 45,341.5 mass is held up by a 101.6-mm diameter steel cylinder surrounded by a copper tube of equal length with an outside diameter of 203.2 mm which fits smugly around the steel cylinder. The weight is distributed uniformly over the surfaces of the supporting members. What is the stress in the copper tube?
E (steel) = 206.85 GPa
E (copper) = 110.32 GPa

A) 10,563 MPa
B) 10,908 MPa
C) 11,253 MPa
D) 11,597 MPa

18.
What is the magnitude of the force acting on member CD?
A) 3,700 N
B) 3,766 N
C) 4,132 N
D) 4,198 N

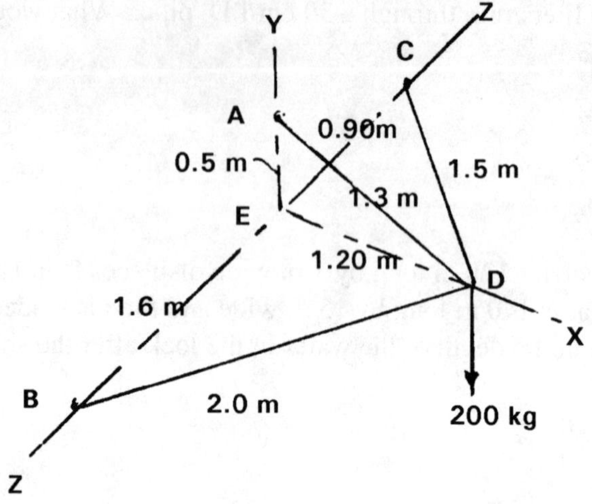

Figure 18.

19.
A 25.0 mm thick steel plate 0.25 m wide is loaded as shown. What is the maximum stress in the plate? See Figure 19.
A) 1.60 MPa
B) 2.91 MPa
C) 4.22 MPa
D) 5.52 MPa

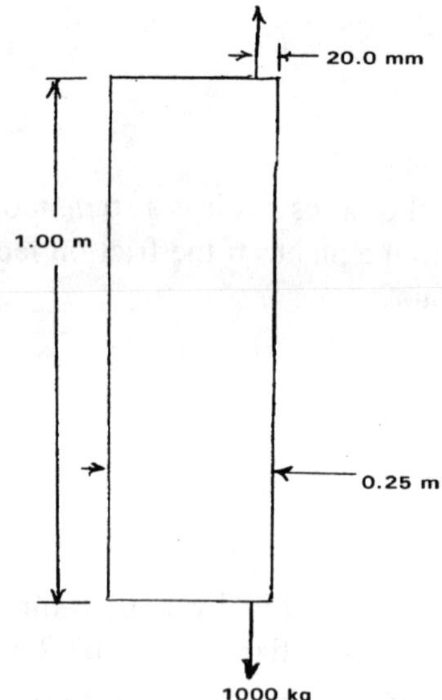

Figure 19.

20.
A closed-end tube 152.4 mm O.D. with a 2.54 mm wall thickness contains a fluid at a pressure of 3.448 MPa. It is subjected to a torsional load of 678 Nm. What is the maximum principal stress in the tube?
A) 53 MPa
B) 67 MPa
C) 93 MPa
D) 101 MPa

21.
From an instrument that is considered to be a secondary standard with an error less than +/- 0.5% and from another instrument with an error less than +/- 1% is calibrated from the secondary standard, statistically what is the probably error of the second instrument?
A) 0.5%
B) 0.75%
C) 0.95%
D) 1.12%

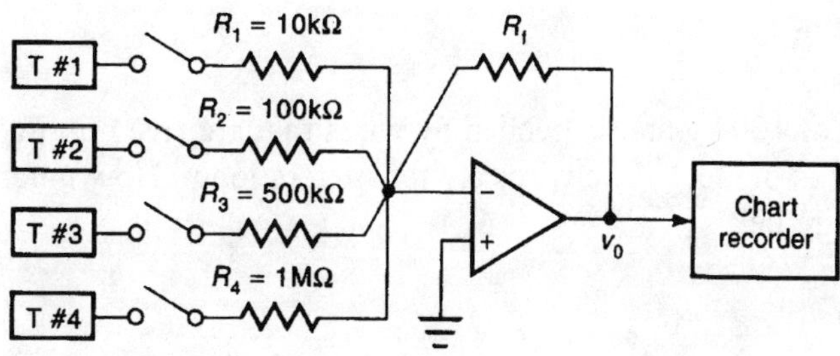

22.
Two different transducers are to be used at different times to drive a strip-chart recorder and the interfacing between the transducers will be an op-amp (see figure 7.2). Assume the op-amp has a 1 megohm resistor in feedback across the op-amp, also assume the full scale expected output of the first transducer will be less than 500 mv and the second one less than 200 mv. If the recorder has a full scale display of 10 volts, what should R_1 and R_2 be (for a full scale display on the recorder)?

A)R1=10kΩ,R2=20kΩ
B)R1=30kΩ,R2=10kΩ
C)R1=50kΩ,R2=20kΩ
D)R1=80kΩ,R2=40kΩ

23.
It is necessary to measure a temperature range from -40° to + 120°C and it is important to know the temperature within a quarter of a degree. A certain linear transducer is capable of producing a variation of voltage over this range. Assume an A/D converter will produce a binary output that will represent the temperature readings. Determine the number of bits (n rating of the A/D converter) needed.
A)4
B)10
C)12
D)16

24.
Why is it not practicable to join aluminum wire to copper wire for an electrical connection by twisting the two wires together?
A) It would not be mechanically strong.
B) It would corrode in time and provide poor conduction.
C) Electrical connections should be soldered.
D) Aluminum wire is not as ductile as copper wire.

25.
Aluminum bolts two cm in diameter and 500 cm long are used to hold the head on a pressure vessel. One bolt has a sensitive extensometer attached to the body of the bolt. It measures an elongation of 0.00400 mm in a ten-cm length of the bolt when it is tightened. How much force is exerted by the bolt?
A) 8,670 N
B) 8,770 N
C) 8,870 N
D) 8,970 N

26.
Two masses of 100 kg are suspended by wires that are five mm in diameter. One wire is of aluminum and the other is of steel. The wires are ten meters long. How much lower will the mass held by the aluminum wire be?
A) 4.37 mm
B) 4.87 mm
C) 5.37 mm
D) 5.87 mm

27.
An aluminum pipe is buried in the earth between two buildings. What would you suggest to prevent excessive corrosion of the pipe?
A) Bury a an excess scrap copper wire alongside the aluminum pipe.
B) Run the aluminum pipe alongside an existing iron pipe water line.
C) Place pieces of junk magnesium in the ditch beside the aluminum pipe.
D) Replace the earth around the pipe with sand.

28.
What would be the expected increase in efficiency of an Otto-cycle gasoline engine if the compression ratio were increased from 6.5 to 8.0?
A) 4.6%
B) 5.9%
C) 7.2%
D) 8.5%

29. A home owner plans to use solar heat to warm his house. He figures to heat water, store it, and then circulate the warm water through coils to heat his living area. The design calls for 5.0 kW of roof heating to supply the reserve heat for the water reservoir. If his solar panels will provide a 20% conversion of incident energy to heat water running through piping attached to the panels, and it is estimated that in his area the radiant solar energy will average 1.35 kW/m^2, how large a roof area must he cover with the solar panels?
A) 18.5 m^2
B) 20.7 m^2
C) 22.9 m^2
D) 25.0 m^2

30. A production line moves in steps of two meters. Between each two steps there is a dwell time of one and a half minutes. During that time it is necessary to install thirty cap screws at one location. Time study tests have shown that an operator can easily locate, thread in, and tighten ten cap screws in one minute. How many operators should be assigned to this station?
A) One.
B) Two.
C) Two and a half (one operator to divide time between two adjacent stations)
D) Three.

31. Five liter containers are to be filled with oil with a specific gravity of 0.85. The contents must be kept within 0.10% of the specified amount. The filling system is controlled by a scale which automatically cuts off the flow of oil when the prescribed amount has been put into the container. What should be the accuracy of the scale?
A) 0.0353 N
B) 0.0417 N
C) 0.0472 N
D) 0.0516 N

32. Containers to hold 20.0 liters of a liquid are filled on a production line which moves in pulses controlled by a geneva motion device. At each stop a nozzle lowers into a container, opening a valve in the nozzle. Then a cylinder filled with the exact amount of the liquid is emptied into the container by a piston inside the cylinder. If it requires one second to insert the nozzle and one second to remove it, at what rate could the containers be filled providing the liquid velocity is limited to 6.00 m/sec and the nozzle has an I.D. of 5.00 cm?
A) 10.2/min
B) 12.2/min
C) 14.2/min
D) 16.2/min

33. What should be the width of a composition belt with thickness of 9.0 mm to deliver 100 kW of power to a driven pulley in a sawmill? The following design criteria control:
1. Belt material ultimate strength, 27.5 MPa.
2. Use a safety factor of ten.
3. Pulley diameters of 60.0 cm (both pulleys)
4. Belt velocity of 1,200 meters per minute.
5. The belt is initially tightened with 4.5 kN tension on each side of the driving pulley.

A) 23 cm
B) 28 cm
C) 33 cm
D) 38 cm

34. What would be the stress in a 2.0 cm diameter shaft that transmitted two kW of power at 1750 RPM?
A) 5.45 MPa
B) 5.95 MPa
C) 6.45 MPa
D) 6.95 MPa

35. What size engine is required to propel a 3-ton vehicle up a 6% grade at 90 km/hr ?
A) 34 kW
B) 39 kW
C) 44 kW
D) 49 kW

36. How much force would a 75 kg pilot exert on his plane seat at the top of a vertical loop which has a radius of 300 meters if the speed of the plane were 300 km/hr ?
A) 1,136 N
B) 1,336 N
C) 1,536 N
D) 1,736 N

37. Two 1.5 kg masses are connected by a massless string hanging over a smooth frictionless peg. A third mass of 1.5 kg is added to one of the other masses. Of the system is released, what is the force on the peg?
A) 44 N
B) 39 N
C) 42 N
D) 37 N

38. What theoretical power is required for the isothermal compression of 25 cubic meters of air per minute from one atmosphere to 830 kPa absolute pressure.
A) 94 kW
B) 89 kW
C) 84 kW
D) 79 kW

39. A rigid container holds 1.5 kg of air at 175 kPaa and 40 C. What is the volume of the container?
A) 0.57 m^3
B) 0.67 m^3
C) 0.77 m^3
D) 0.87 m^3

40. Heat is lost through one m² of furnace wall at the rate of 1.639 kW. The temperature on the inside of the furnace wall is 1040 C, and it is 0.500 m thick. The average thermal conductivity of the furnace wall is--1.056 W/m·K). What is the temperature of the outside surface of the wall?

A) 264 C
B) 269 C
C) 274 C
D) 279 C

41. A counterflow heat exchanger operates with hot liquid entering at 200 C and leaving at 160 C. The liquid to be heated enters at 40 C and leaves at 140 C. What is the log mean temperature difference between the hot and cold liquids?

A) 87 C
B) 92 C
C) 97 C
D) 102 C

42. A beaker of liquid is set on a bench to cool. If the temperature of the liquid falls from 70 C to 60 C in 12 minutes , how much longer will it take for the liquid to cool to 50 C? The ambient temperature is 20 C. Neglect extraneous losses.

A) 13.5 min
B) 14.2 min
C) 14.9 min
D) 15.5 min

43. The wall of a house is made up of a layer of plaster over brick. The coefficients of conductance are as follows:

Film coefficient for indoor surface	9.37 W/m²K
Plaster layer	26.35 W/m²K
Brick outer layer	3.58 W/m²K
Film coefficient for outer surface	34.07 W/m²K

What is the overall heat transfer coefficient for the wall?

A) 2.21 W/m²K
B) 1.84 W/m²K
C) 1.46 W/m²K
D) 1.32 W/m²K

44. For the conditions in the above problem, what would be the rate of heat flow through 14 sq. meters of wall area if the outside temperature were -2.0 C and the inside temperature were 21 C?

A) 650 W
B) 710 W
C) 770 W
D) 810 W

45. Air in the amount of 280 m³ per minute is to be heated from 20 C to 75 C in a tubular heater by means of saturated steam at 107 C condensing on the outside of the tubes. How much steam would be required if the heat of vaporization of steam at 107 C is 2.24 MJ/kg ?

A) 423 kg/hr
B) 448 kg/hr
C) 473 kg/hr
D) 498 kg/hr

46. A 15 cm diameter cylinder has a mass of 1.10 kg. It is placed in a cylindrical barrel which is 30 cm inside diameter and is partly full of water. The inserted cylinder floats. How much will the surface of the water in the barrel rise?

A) zero
B) 0.82 cm
C) 1.56 cm
D) 2.10 cm

47. A jet of water 2.5 cm in diameter flows at the rate of 0.500 m³/min. It impinges on a flat surface perpendicular to the axis of the stream. What force is exerted on the surface?

A) 142 N
B) 147 N
C) 152 N
D) 157 N

48. A steel band is shrunk onto a 30 cm diameter aluminum cylinder. The band is 2.00 mm thick and the diametral interference fit is 0.0108 mm. What is the stress in the steel band?

A) 6.56 MPa
B) 7.56 MPa
C) 8.56 Mpa
D) 9.56 MPa

49. A cable extends upward from a post at an angle of 30° with the vertical. It passes over a pulley at a point two meters away horizontally (3.464 m higher) and supports a mass of 900 kg. The post is two meters high (from the ground to the point of attachment of the cable) and is set in concrete. The section modulus of the post is 78.64 cm³, and its cross-sectional area equals 18.5 cm². What is the maximum stress in the pole?

A) 108 MPa
B) 112 MPa
C) 116 MPa
D) 120 MPa

GO TO THE NEXT PAGE

50. A distance of 1,579.560 meters was measured with a steel tape on a cold day at -5.5 C. The tape was calibrated at 20 C. The coefficient of thermal expansion for steel is 11.7×10^{-6}/C. What was the true distance?

A) 1,579.560 m
B) 1,579.089 m
C) 1,581.321 m
D) 1,580.031 m

51. A pressure gage on a tank reads 689.64 kPa. The gage was calibrated on a dead-weight gage tester and found to read 0.025% low over the range from 600 to 700 kPa. What was the true pressure in the tank?

A) 688.47 kP
B) 689.47 kPa
C) 689.64 kPa
D) 689.81 kPa

52. A U-tube mercury manometer is to be used to measure an estimated differential pressure of 88 cm of water in a water system. Using a safety factor of 1.5, how long should the measurable section of the U-tube be?

A) 8.33 cm
B) 9.04 cm
C) 9.71 cm
D) 10.47 cm

53. A cantilever beam 5.0 cm wide by 10.0 cm high and 3.0 m long holds a 700 kg load on its end. What would be the percentage increase in the deflection if the beam were made of aluminum rather than steel?

A) 194%
B) 204%
C) 254%
D) 304%

54. A diesel-engine-driven generator produces 10.0 kW of power. The generator has an efficiency of 96%. If the diesel engine operates with an efficiency of 34%, how many liters of fuel would it consume in an eight-hour shift?
Diesel fuel has a specific gravity of 0.81 and a heating value of 42.57 MJ/kg.

A) 26ℓ
B) 30ℓ
C) 41ℓ
D) 35ℓ

55. Outside air at 5 C is heated to 25 C before it is introduced into an occupied room. If the relative humidity of the outside air is 70% what would be the relative humidity, RH, of the heated air?

A) 65%
B) 42%
C) 31%
D) 18%

56. How much moisture is contained in one m^3 of dry air if the dry bulb temperature is 29 C and the wet bulb temperature measures 20 C?

A) 9.6 g
B) 11.5 g
C) 13.5 g
D) 15.6 g

57. A refrigerant leaves the condenser as a liquid at 26 C and expands into the evaporator which is at 1.5 C. Data from a table show a refrigerating effect of 130.85 J/kg. What would be the mass flow rate for a refrigeration requirement of 52.75 kW?

A) 20.72 kg/min
B) 22.09 kg/min
C) 23.17 kg/min
D) 24.19 kg/min

58. What size motor would be required to operate a fan which is to supply 30.0 m^3/min of air against a static pressure of 12.0 cm of water pressure if the fan efficiency is 78%? Assume standard conditions of one atmosphere and 20 C.

A) 654 W
B) 694 W
C) 754 W
D) 794 W

59. What size motor would be required to pump 350 ℓ/min of fluid with a density of 1.20 against a pressure of 360 kPa if the efficiency of the pump is 85%?

A) 2.25 kW
B) 2.50 kW
C) 2.75 kW
D) 3.00 kW

60. A centrifugal pump delivers 1,200 ℓ/min of water at a discharge pressure of 165 kPa and a suction pressure of 5.0 cm of mercury. The motor draws 6.30 kW of power. What is the overall efficiency of the motor-pump combination?

A) 55%
B) 60%
C) 65%
D) 70%

End of Exam. Check your work.

Sample Exam Solutions

1.

Watt = Joule/sec = Nm/sec so 4,026.8 kW = 4.0268 MNm/sec

1,200 rpm = 20 rps, and a point 0.1524 m out from the axis of the shaft would travel at a velocity

$$v = 20 \times 2\pi \times 0.1524 = 19.15 \text{ m/s}$$

The cross-sectional are of a bolt would equal

$$0.0254^2 \times \pi/4 = 506.7 \times 10^{-6} \text{ m}^2$$

Allowable force on one bolt =

$$506.7 \times 10^{-6} \times 103.4 \text{ MN/m}^2 = 52.393 \text{ kN/bolt}$$

(4.0268 MNm/s)/(19.15 m/s) = 0.2103 MN at bolt circle

210,300/52,393 = 4.01 bolts required, so use four bolts. **Answer is C).**

2.

Stress in the outer fibers of a shaft subject to torsion = Tc/J'

where $J' = \pi \times \text{Diam}^4/32$

110.32 MPa = (33.900 kJ × D/2)/J'

D/J' = 6,508.55

3° angle of twist = 3 × π/180 = 0.05236 radian

$$\theta = T\ell/(GJ') = 33.900 \text{ kJ} \times 2.54 \text{ m}/(82.740 \text{ GPa} \times J')$$

$$0.05236 = 1.0407 \times 10^{-6}/J'$$

$$J' = 1.9876 \times 10^{-5}$$

$$D = 6{,}508.55 \times 1.9876 \times 10^{-5} = 0.1294 \text{ m}$$

The diameter of the bore can be determined with the aid of the relationship for the stress Tc/J'

where $J' = \pi(D^4 - d^4)/32 =$

$$J' = 1.9876 \times 10^{-5} = (\pi/32) \times (D^4 - d^4)$$

$$20.2455 \times 10^{-5} = 24.037 \times 10^{-5} - d^4$$

$$d = (7.719 \times 10^{-5})^{-1/4} = 0.0937 \text{ m}$$

D/d = 1.381 **Answer is D).**

3.

Each bolt will withstand one-quarter of the vertical load "F" plus the force applied due to the moment resulting from the applied load. The applied moment will equal 0.381 Nm where 0.381 meter is the distance from the point of application of the force perpendicular to the horizontal distance to the centroid of the bolt pattern. The moment will be resisted equally by the four bolts, and the force exerted on a bolt will be perpendicular to the line from the bolt to the centroid. For bolts "A" and "B" the applied force will have a component in the same direction as the applied force, so the total loads on bolts "A" and "B" will be greater than the total loads on bolts "C" and "D". The moment force applied to each bolt will equal 0.381F/(4×0.127) = 0.750F Where 0.127 m is the distance of the moment arm to a bolt from the centroid of the bolt pattern. The moment-producing force acting on bolt "A" will act downward at an angle of 36.87° from the vertical. The two components of this force wil equal--

$$F_y = 0.80 \times 0.75 \text{ F} = 0.60 \text{ F}$$

$$F_x = 0.60 \times 0.75 \text{ F} = 0.45 \text{ F}$$

The total vertical force acting on bolt "A" would equal--

0.25F + 0.60 F = 0.85 F and horizontal force = 0.45 F

total force acting on bolt "A"

$$= F \times \sqrt{(0.85^2 + 0.45^2)} = 0.9618 \text{ F}$$

So F = 8,896/0.9618 = 9,249 N **Answer is B).**

4.

The maximum load which can be held by the rod equals--

$$15.75^2 \times 10^{-6} \times \pi/4 \times 137.9 \times 10^6 = 26{,}867 \text{ N}$$

The area of the washer contacting the wood must equal--

$26{,}879/(2.379 \times 10^6) = 0.011298\ m^2$

The washer will have a 19.05-mm diameter hole in its center so the outer diameter of the washer would equal--

$(0.011298 + 19.05^2 \times \pi/4 \times 10^{-6})^{1/2}/0.7854 = 0.1370\ m = 137\ mm$ **Answer is D).**

5.

The velocity of the surface of the drum--

vel = $1{,}250/60 \times \pi \times 0.254 = 16.624$ m/s

Frictional force exerted by brake shoe on surface of drum--

F = $111.20 \times 0.21 = 23.352$ N

Power dissipated = $23.352 \times 16.624 = 388.20$ J/s or 388.20 watts. **Answer is A).**

6.

Force perpendicular to track exerted by car = kgf × cos 5°

$68{,}027 \times 0.996195 = 67.768$ kgf = 664.57 kN

Friction resisting force = $0.05 \times 664.57 = 33.229$ kN

Gravitational force acting on car parallel to track

F = 68,027 kgf × sin 5° = $68{,}027 \times 0.087156 = 5{,}929$ kgf or 58.143 kN

Net force acting on car parallel to track = 58.143 kN - 33.229 = 24.914 kN

This force acts over a distance of 304.8 m producing a total amount of energy at impact of 7.5938 MJ

Energy absorbed = $\int Fds$ where force exerted by spring = ks

energy absorbed = $\frac{1}{2}ks^2 = \frac{1}{2}k \times 1.0668^2 = 0.56903k$

so k = 7.5938×10^6 Nm/0.56903 = 13.345 MN/m. **Answer is A).**

7.

Mass of disk = $(\pi/4) \times 50.0^2 \times 2.0 \times 7.87 = 30.91$ kg

The moment of inertia of circulsr disk about an axis through its center and perpendicular to the disk = $\frac{1}{2}MR^2$

$I = \frac{1}{2}MR^2 = \frac{1}{2} \times 30.91 \times 0.25^2 = 0.9659$ kg-m²

$\omega = \sqrt{(GJ/IL)}$

Where: G = shear modulus of the steel rod = 8.3×10^{10} Pa

J = Polar moment of inertia of the rod = $\frac{1}{2}\pi r^4$

$J = \frac{1}{2}\pi \times 0.0050^4 = 9.818 \times 10^{-10}$

I = Moment of inertia of the disk = 0.9659

$\omega = \sqrt{([8.3 \times 10^{10} \times 9.818 \times 10^{-10})/(0.9659 \times 1.00)]} = \sqrt{84.37}$

$\omega = 9.186$/sec angular frequency

$f = \omega/(2\pi) = 1.462$ $\tau = 1/f = 0.684$ sec. **Answer is B).**

8.

Net force acting on the cart to cause motion--

2T to the left resisted by the inertial or d/Alembert force of $m_c a$

Net Force = 2T - 500a = 0

giving 2T = 500a

The force applied by the suspended mass--

Force down = $m_s g$ resisted by 2T + $m_s a$

Or 2T = 500×9.8066 - 500a but 2T = 500a

so 1,000a = 4.9033 kN and a = 4.9033 m/sec²

T = $(500/2) \times 4.9033 = 1{,}225.8$ N or 1.23 kN **Answer is B).**

9.

Work = $(P_2V_2 - P_1V_1)/(1 - k) = 63{,}300$ J

$63{,}300 \times (-0.40) = P_2V_2 - 103{,}425$ Pa $\times 0.2831\ m^3$

$P_2V_2 = 29{,}280 + 25{,}320 = 54{,}600$ J

$P_2V_2 = m_g RT_2 = m_g \times 286.8 \times (190.6 + 273.2) = m_g \times 133{,}018$

where m_g = mass of gas

$T_1 = P_1V_1/m_gR = 103{,}425 \times 0.2831/(0.410 \times 286.32) = 249.0$ K
$T_2 = 190.6 + 273.2 = 463.8$ K
$\Delta T = 463.8 - 249.0 = 214.8$ K
$\Delta U = m_g c_v \Delta T = 0.410 \times 718 \times 214.8 = 63{,}233$ J **Answer is B).**

10.
$PV = m_gRT$
V = constant = 1.416 m³ and R constant
where m_g = mass of gas
Differentiate and divide by dt
$V\, dP/dt = Rm_g\, dT/dt + RT\, dm_g/dt$ mass of gas at time zero
$m_g = PV/RT = 1.379 \times 10^6 \times 1.416/(286.7 \times 388.7) = 17.52$ kg
$V\, dP/dt = 1.416 \times 137.9$ kPa/sec
$= 286.7$ J/kgK $\times 17.52 \times 27.78 + 286.7 \times 388.7 \times dm_g/dt$
$dm_g/dt = 0.500$ kg/sec **Answer is D).**

11.
Use the general energy equation:
For steady state (turbines etc.) for one kg--
$h_1 + Vel_1^2/2 + q = h_2 + Vel_2^2/2 + Work$ but $h = u + Pv$
$u_1 + P_1v_1 + Vel_1^2/2 + q = u_2 + P_2v_2 + Vel_2^2/2 + Work$
$Vel_1^2/2 = 152.4^2/2 = 11{,}613$ J/kg
$P_1v_1 = 689{,}500 \times 0.250 = 172{,}375$ J/kg
Energy in:
$11{,}613 + 172{,}375 - 23{,}263 + u_1 = 160{,}725 + u_1$ J/kg
$Vel_2^2/2 = 304.79^2/2 = 46{,}448$ J/kg
$P_2v_2 = 137{,}900 \times 0.9365 = 129{,}143$ J/kg
Energy out:
$46{,}448 + 129{,}143 + 582{,}938 + u_2 = 758{,}529 + u_2$ J/kg
$u_1 - u_2 = -597{,}804$ J/kg **Answer is A).**

12.
For steady state
$h_1 + Vel_1^2/2 = h_2 + Vel_2^2/2$
$\Delta h = c_p\, \Delta T$
$c_p = c_v \times k = 1.687 \times 1.30 = 2.1931$ kJ/kg○K
$T_2 = T_1 \times (P_2/P_1)^{(k-1)/k} = (26.67 + 273) \times (551.6/689.5)^{0.2308}$
$T_2 = 299.67 \times 0.800^{0.2308} = 284.63$K
$\Delta T = 299.67 - 284.63 = 15.04$
$c_p \times \Delta T = 2.1931 \times 15.04 = 32.98$ kJ/kg○K
$\Delta h = 2.1931 \times 15.04 = 32.980$ kJ/kg○K
$Vel_2^2/2 - Vel_1^2/2 = \Delta h = 32{,}980$
$Vel_2^2 = 121.9^2 + (2 \times 32{,}980) = 80{,}820$
$Vel_2 = 284$ m/sec. **Answer is D).**

13.
The friction factor will be a function of the Reynolds number. $Re = \rho Dv/\mu$
$\rho = 0.94 \times 1{,}000 = 940$ kg/m³
D = 0.300 m A = 0.707 m²
$v = (2{,}000/1{,}000)/(0.707 \times 60) = 0.0471$ m/s
$\mu = 201 \times 0.001 = 0.201$ Pa·s
$Re = 940 \times 0.300 \times 0.0471/(0.201/g) = 648$

This indicates that the flow will be well into the laminar-flow range so f = 64/Re = 64/648 = 0.0988
Answer is D).

14.
The ship displaces 8,500 m³ of fresh water before loading. This would cause a rise of 8.500/(140 × 15) = 4.0476 m in the lock. The loaded ship would displace--
8,500 + 3,500 = 12,000 m³ of water and would cause a total rise of 12,000/2,100 = 5.714 m in the lock, so the depth would increase by 1.667 m. **Answer is B).**

15.
The velocity of the water out of the tank will equal--
$v = \sqrt{(2gh)} = \sqrt{(2 \cdot 9.8066 \cdot 9.00)} = 13.286$ m/s
The jet is 1.00 m above the ground. $h = v_0 t\ \tfrac{1}{2}at^2$
For this case the vertical velocity at time zero is zero. The time for the water to fall 1.00 m
$t = \sqrt{(2 \cdot h/g)} = 0.452$s The distance from the base of the tank the water will strike--
$s = 0.452 \times 13.286 = 6.00$ m **Answer is C).**

16.
Apply Bernoulli's equation between the surface of the reservoir and the discharge from the pipe. There is no energy added and the velocity of the surface of the water is zero. The pressure at the surface and at the pipe discharge are both atmospheric, so they cancel. The only term remaining on the left side of the equation is the difference in elevation between the surface of the reservoir and the outlet of the pipe, $Z_1 - Z_2 = 150$ m. On the right hand side of the equation would be the velocity term $v^2/2g$ and the work done in overcoming pipe friction--
$h_L = fL/D \times v^2/2g$ $fL/D = 0.02 \times 200/0.30 \times v^2/2g$
so $150 = (1 + 13.33) \times v^2/2g$ and $v = 14.33$ m/s
$Q = 14.33 \times 0.0707 = 1.013$ m³/s **Answer is B).**

17.
The deflections for the steel cylinder and the copper tube will be equal and since the lengths are the same, the unit deflections, δ, will also be equal.
Stress = Eδ
Steel area = $101.6^2 \times \pi/4 = 8{,}107$ mm² = 0.008107 m²
Force exerted by steel cylinder = $0.00810734 \times E_s \times \delta$
Copper area = $(203.2^2 - 101.6^2) \times \pi/4 = 24{,}322$ mm² = 0.024322 m²
Force exerted by mass F = 45,341.5 × 9.8066 = 444.717 kN
Force exerted by copper tube = $0.024322 \times E_c \times \delta$
$444{,}717 = 0.008107 \times 206.85 \times 10^9 \times \delta +$
$0.024322 \times 110.32 \times 10^9 \times \delta = 4.3601 \times 10^9 \times \delta$
$\delta = 444{,}717/(4.3601 \times 10^9) = 101{,}997 \times 10^{-9}$ m/m
Stress in copper tube =
$101{,}997 \times 10^{-9} \times 110.32 \times 10^9 = 11{,}253$ Mpa **Answer is C).**

18.
The force exerted by a 200 kg mass = 200 × 9.8066 = 1,961 N
Force in member AD = 13/5 × 1,961 = 5,099 N
Force in member ED = X component of force in AD
= 1.2/1.3 ×5,099 = 4,707 N
The Z components of forces in members CD and BD in the X-Z plane are equal
$(0.9/1.5) \times F_{CD} = (1.6/2.0) \times F_{BD}$
so Force in BD = 0.75 × Force in CD
The sum of the X components of forces CD and BD equal the X component of the force in AD

$(1.2/1.5) \times F_{CD} + (1.2/2.0) \times F_{BD} = (1.2/1.3) \times F_{AD}$
0.800 CD + 0.600 × 0.750 × CD = 4,707 N
Force in CD = 4,707/1.250 = 3,765.6 N **Answer is B).**

19.
The maximum stress will occur at the edge of the plate and will be made up of the direct tensile stress due to the applied load plus the bending stress due to the eccentricity of the load.
F/A = 1,000 × 9.8066/(0.250 × 0.025) = 1.569 MPa
The stress due to the eccentric load S = Mc/I
M = (0.125 - 0.020) × 9,806.6 = 1,029.7 Nm or 1,029.7 J
c = 0.250/2 = 0.125
$I = bh^3/12 = 0.025 \times 0.250^3/12 = 32.55\ \mu m^4$
Bending stress = $1,029.7 \times 0.125/(32.55 \times 10^{-6}) = 3.954$ MPa
The maximum tensile stress = 1.569 + 3.954 = 5.523 MPa **Answer is D).**

20.
The maximum prncipal stress will be calculated using the method of Mohr's Circle as given in the FE Reference Handbook supplied by the NCEES.
Assume a thin-walled tube then $\sigma = PD/2t$
The tube I.D. = 152.4 - 5.08 = 147.32 mm or 0.14732 m
Hoop stress--
$\sigma_y = (3.448 \times 10^6 \times 0.14732)/0.00508 = 99.992$ MPa
Longitudinal stress, $\sigma_x = P \times A_{ID}$/Area of tube wall
$A_{ID} = \pi/4 \times 0.14732^2 = 0.017046\ m^2$
Area of tube wall = $\pi/4 \times (O.D.^2 - I.D.^2) = 0.001195\ m^2$
$\sigma_x = (3.448 \times 10^6 \times 0.017046)/0.001195 = 49.184$ MPa
Shear stress $s_s = Tc/J$ where $J = \pi/32 \times (O.D.^4 - I.D.^4)$
$J = 6.7152 \times 10^{-6}\ m^4$
$s_s = 678 \times 0.0762/6.7152\times10^{-6} = 7.6935$ MPa
From the figure of Mohr's Circle in the FE Handbook
$\tau_{max} = \sqrt{[s_s^2 + \{(\sigma_x - \sigma_y)/2\}^2]} = \sqrt{(7.6935^2 + 25.404^2)}$
=26.45 MPa
$\sigma_1 = (\sigma_x + \sigma_y)/2 + \tau_{max} = 74.588 + 26.45 = 101$ MPa **Answer is D).**

21.
From equation 7.1

$$e_c = \sqrt{e_1^2 + e_2^2 + e_3^2 ... e_n^2}$$

$e_c = [(0.5)^2 + (1)^2]^{1/2} = 1.12,$

therefore, statistically, the error is less than +/- 1.12%. **Answer is D).**

22.
The gain needed for the first transducer is 20 and the second is 50, therefore (ignoring the minus sign of the inverting circuit), the equations are,

$Gain_1 = R_f/R_1 = 20,\ R_1 = 1x10^6/20 = 50$ k ohms,

$Gain_2 = R_f/R_2 = 50,\ R_1 = 1x10^6/50 = 20$ k ohms. **Answer is C).**

23.

The number of increments or resolution must be at least 160x4=640. Therefore $2^n > 640$. For an n= 9, gives 512, while an n=10 gives 1024. Therefore an n of 10 is required.

Answer is B).

24.

High mechanical strength is not needed in an electrical connection. Soldering is not necessary for low-current applications when two copper wires are connected and joined with a twist-on connector, and aluminum cannot be soldered to copper. The ductility of the conductors is not important in an electrical connection. From the table of oxidation potentials in the FE Handbook it is seen that aluminum has a much higher potential than copper. Thus it would form the anode of a galvanic cell and corrode. Due to the large potential difference the corrosion would proceed rapidly when an electrolyte formed from the moisture in the air. It might also be noted that the corrosion would result in a poor connection, and would result in a high resistance in the connection which could result in a fire.

The correct answer is B).

25.

Measured strain is--

$\epsilon = 0.00400/10 = 0.000400$ cm/cm

Stress $= \epsilon \cdot E = 0.000400 \times 6.9 \times 10^{10} = 27.60$ MPa

Area of bolt = 0.000,3142 m²

$F = S \times A = 0.3142 \times 27.6$ kPa = 8,672 N

The correct answer is A).

26.

The stresses in the two wires will be equal.

$S = 980.66/(1.9635 \times 10_{-5}) = 50.00$ MPa

$\epsilon_S = 50.00 \times 10^6/(2.1 \times 10^{11}) = 238.1 \times 10^{-6}$ m/m

ΔL for steel wire = 2.381 mm

$\epsilon_{Al} = 50.00 \times 10^6/(6.9 \times 10^{10}) = 724.6 \times 10^{-6}$ m/m

ΔL for aluminum wire = 7.246 mm

The mass held by the aluminum wire will be 4.87 mm below the mass held by the steel wire.

The correct answer is B).

27.

A study of the table of oxidation potentials in the FE Handbook shows that aluminum has a higher oxidation potential than copper or iron, so B), and especially A), would only accelerate the corrosion of the aluminum pipe. Magnesium has a higher oxidation potential so it would act as a sacrificial anode and protect the aluminum pipe. Selection D) would do nothing to protect the aluminum pipe.

The correct answer is C).

28.

From the FE Handbook the efficiency of an Otto-cycle engine is given as $\eta = 1 - r^{1-k}$ where r equals the compression ratio.

For r = 6.5

$$\eta = 1 - 6.5^{-0.4} = 1 - 0.473 = 0.527 \text{ or } 52.7\%$$

for r = 8.0

$$\eta = 1 - 8.0^{-0.4} = 1 - 0.435 = 0.565 \text{ or } 56.5\%$$

There would be an increase of 3.80 percentage points or an increase of 7.21% in the engine efficiency.

The correct answer is C).

29.
The energy available for heating the water equals--
$0.20 \times 1.35 = 0.270$ kW/m² of solar paneling
$5.00/0.270 = 18.5$ m²
The correct answer is A).

30. If one operator can easily handle ten cap screws in one minute, then one operator can handle 15 in a minute and a half. Two operators could then handle the requirement of 30 in a minute and a half.
The correct answer is B).

31. The nominal mass of oil would equal--
$5 \times 1.00 \times 0.85 = 4.250$ kg which would produce a force of 41.68 N For a 0.10% accuracy the weighing device would have to be accurate to within 0.0417 N.
The correct answer is B).

32. The rate of flow through the nozzle equals--
$6.00 \times 0.001964 = 0.0118$ m³/sec or 11.8 ℓ/sec
Filling time equals $20/11.8 = 1.695$ sec
To this would have to be added 2.0 sec insertion and removal time giving a total of 3.695 sec, giving a theoretical rate of $60/3.695 = 16.24$ per minute.
The correct answer is D).

33.
The maximum safe design stress equals 2.75 MPa. Assume the increase in belt tension on one side of the pulley equals the decrease in tension on the other side.
Power $= Tq \times \omega = (F_1 - F_2)r \times \omega$ $\omega = v/r$
$P = (F_1 - F_2) \times v$
$[(4.5 + \Delta) - (4.5 - \Delta)]$ kN $\times$ 20 m/s = 100 kW $\Delta = 2{,}500$ N
Maximum required tension in belt = 7.00 kN
Required belt area = 7.00/2,750 = 0.00255 m²
Thickness = 0.0090 m so width = 28.3 cm
The correct answer is B).

34. Two kW equals 2,000 N·m/s
1,750 RPM/60 = 29.167 rev/sec
Torque $= 2{,}000/(2\pi \times 29.167) = 10.91$ N·m or 10.91 J
Stress = T·c/(Polar Moment of Inertia) or
$S = 2T/(\pi \cdot r^3) = 6.946$ MPa
The correct answer is D)

35. The car moves at the rate of 25 m/s. A 6% grade gives an angle of 3.434° so when the car travels 100.18 meters on the road it rises 6.00 m. The engine must produce energy in the amount of--
$25 \times 6/100.18 \times 3{,}000$ kg $\times$ 9.8066 N/kg = 44,050 W
The power requirement is 44.05 kW
The correct answer is C)

36. Centripetal force equals $m \cdot v^2/r$
Force $= 75 \times 83.33^2/300 = 1{,}736$ N
The correct answer is D)

37. Initially the two masses are at rest and the tension in the cord equals $1.5 \times 9.8066 = 14.71$ N so the force acting on the peg equals 29.42 N. When the extra 1.5 kg is added to one of the masses and the system is released the larger mass will accelerate downward and the smaller mass will accelerate upward. The tension in the string will equal--

$T = 3.00 \cdot g - 3.00 \cdot a = 1.5 \cdot g + 1.5 \cdot a$

$a = 3.2689$ m/s²

$T = 19.613$ N force on peg = 2T = 39.23 N

The correct answer is B).

38. The gas law states $PV = mRT$ if the temperature is constant then PV = constant.

Work = $\int P \cdot dV$ which equals, for an isothermal process--

$W = PV \ln(V_2/V_1)$ work done by the gas

For constant T $V_2/V_1 = P_1/P_2$

For work done on the gas--

$W = 101.3 \text{ kPa} \times 25 \times \ln(830/101.3) = 5{,}327$ kJ/min

Power required = 5,327,000/60 = 88,783 watts or 88.8 kW

The correct answer is B)

39. Assume air is a perfect gas, then--

$PV = mRT$ $R_{air} = 8{,}314/29{,}000 = (0.2867/K)$joules

$V = mRT/P = 1.5 \times 0.2867 \times 313/175 = 0.7692$ m³

The correct answer is C).

40. The rate of heat flow equals--

$Q = k \cdot A \cdot \Delta T/\Delta L$ $1{,}639 = 1.056 \times 1.0 \text{ m}^2 \times \Delta T/0.500$

$\Delta T = 776$ C Outside wall temperature equals--

$T = 1040 - 776 = 264$ C

This could also be calculated as--

5,900,000 MJ/hr = 3,800 (J/hr-m-K) × 1.0 m² × $\Delta T/0.500$

giving $\Delta T = 776$ C as before

The correct answer is A).

41. Log mean temperature difference for a counterflow heat exchanger equals:

$(\max \Delta T - \min \Delta T)/\ln(\max \Delta T/\min \Delta T)$

max ΔT equals hot leaving temperature minus cold entering temperature. For this case--

$\max \Delta T = 160 - 40 = 120$ C

min ΔT equals hot entering temperature minus heated liquid leaving temperature. For this case--

$\min \Delta T = 200 - 140 = 60$ C

log mean temperature difference =

$(120 - 60)/\ln(120/60) = 86.6$ C

The correct answer is A).

42. Newton's law of cooling is expressed as-- $dT/dt = -k \times (T - T_{amb})$

Where T = temperature and t = time

which gives $\int dT/(T - T_{amb}) = -k \int dt$ from T_o to T

$\ln(T - T_{amb})/(T_o - T_{amb}) = -kt$ so--

$e^{-k \cdot 12} = (60 - 20)/(70 - 20) = 0.800$

which gives $k = 0.0186$ per minute

$e^{-0.0186t} = (50 - 20)/60 - 20) = 0.750$

which gives $t = 15.5$ min

The correct answer is D).

43.

$1/U = 1/9.37 + 1/26.35 + 1/3.58 + 1/34.07 = 0.4534$

$U = 2.206$ W/m²K

The correct answer is A).

44. The rate of heat flow--

$Q = U \times A \times \Delta T = 2.206 \times 14 \times 23 = 710.33$ W

The correct answer is B).

45. The mass of air wold equal--

$m = PV/RT = 101.3 \times 280/[(8.314/29) \times 293] = 337.7$ kg/min

$Q = 337 \times 554 \times 1.00 = 18{,}574$ kJ/min

$18{,}574{,}000 \times 60/2{,}240{,}000 = 498$ kg/hr

The correct answer is D).

46. The floating cylinder will displace 1.10 kg of water, or 1,100 cm^3, since the density of water is one gram per cm^3. The inside area of the barrel is 706.9 cm^2. So the surface of the water would rise--

$1{,}100/706.9 = 1.56$ cm.

The correct answer is C).

47. The velocity of the jet equals--

$v = (500{,}000/60)/(2.5^2 \times \pi/4) = 1{,}698$ cm/sec

mass flow rate = 8.333 kg/sec

$F = ma = 8.333 \times 16.98$ m/s $= 141.5$ N

The correct answer is A).

48. The circumference equals pi times the diameter so the circumferential strain in the steel band equals $0.00108 \times \pi/(30 \times \pi) = 3.600 \times 10^{-5}$ m/m or $\epsilon = 0.00108/30 = 3.600 \times 10^{-5}$

Stress $= e\epsilon = 3.6 \times 10^{-5} \times 2.1 \times 10^{11} = 7.56$ MPa

The correct answer is B).

49. The tension in the cable will equal--

$T = 900 \times 9.8066 = 8{,}826$ N. This will apply a lateral force of 4,413 N on the post and a vertical force of 7,644 N. The lateral force will produce a moment of 8,826 Nm at the base of the post. The section modulus, Z, equals I/c. Bending stress:

$S = M/Z = 4{,}413 \times 2/(78.64 \times 10^{-6}) = 112.23$ MPa

to this must be added the tensile stress due to the vertical component of the applied force--

$S = 7{,}644/(18.5 \times 10^{-4}) = 4{,}132$ kPa

Max. tensile stress $= 112.23 + 4.13 = 116.36$ kPa

The correct answer is C).

50. The tape shrank so the distance indicated by the tape was longer than the actual distance. The shrinkage equalled--

$\Delta L = 25.5 \times 11.7 \times 10^{-6} \times 1{,}579.56 = 0.4713$ m the true distance thus equalled $1{,}579.560 + 0.4713 = 1{,}580.031$ m

The correct answer is D).

51. Since the gage reads low, the actual pressure will be higher than the indicated pressure. The error equals 0.00025×689.64 kPa $= 0.172$ kPa. The true pressure equals $689.64 + 0.172 = 689.81$ kPa.

The correct answer is D).

52. The density of mercury from the FE Handbook is 13,560 kg/m^3, or 13.6 times that of water. Since water will be in one leg of the U-tube, the differential pressure will equal 12.6 times the difference in the height of the mercury above that of the water. The difference in the heights of the two columns should equal--

$\Delta H = 88/12.6 = 6.98$ cm

For a safety factor of 1.5 the differential $\Delta H = 10.47$ cm

The correct answer is D).

53. The maximum deflection equals--

$\Delta H = PL^3/3EI$ The only change in the two beams would be the material. The deflection of a steel beam would equal-- Constant/(21×10^{10})

ΔH_{Al} would equal Constant/(6.9×10^{10})

An aluminum beam would deflect 3.043 times as much or 2.043 times more. The deflection would increase 204 percent.

The correct answer is B).

54. The net efficiency of the generating system is $0.96 \times 0.34 = 0.3264$
The heat consumed would equal-- (10,000 J/s)/0.3264 = 30.637 kJ/s
The diesel fuel contains--

$0.81 \times 42.57 = 34.482$ MJ/ℓ of energy

Vol = $30{,}637 \times 3{,}600 \times 8/34{,}482{,}000 = 25.59$ℓ

The correct answer is A).

55. On the psychrometric chart in the FE Handbook locate the intersection of the 5 C temperature line and the 70% RH line. Follow the horizontal line showing moisture content to where it crosses the 25 C temperature line and read the RH at 25 C. The RH at 25 C equals 18%.

The correct answer is D).

56. From the psychrometric diagram in the FE Handbook determine the intersection of the 29 C dry bulb temperature and the 20 C wet bulb temperature. Follow the horizontal line of saturation temperature to the right and read 11 grams of moisture per kg of dry air for the dew point temperature of 15 C. Dry air at 15 C and one atmosphere--

$m = PV/RT = 101{,}300 \times 1.00/[(8{,}314/29) \times 288] = 1.227$ kg/m³

mass of moisture = $11 \times 1.227 = 13.5$ grams

The correct answer is C).

57. The refrigeration requirement is 52.75 kW or 3,165 J/min. The mass rate of flow required would equal 3,165/130.85 = 24.19 kg/min.

The correct answer is D).

58. The density of air equals--

$\rho_{air} = PV/RT = 101{,}300 \times 1/[(8{,}314/29) \times 293)]$

$= 1.206$ kg/m³ and the flow rate 36.18 kg/min

$\Delta P = (1{,}000/1.206) \times 0.12 = 99.50$ m

Power = $(36.18/60) \times 9.8066 \times 99.50 = 588$ W

Motor requirement = 588/0.78 = 754 W

The correct answer is C).

59. The mass rate of flow would be-- $350 \times 1.20 \times 1.0/60 = 7.00$ kg/sec
The head against which the pump would act--

$\Delta H = 360{,}000/[(1{,}200 \text{ kg/m}^3) \times 9.8066] = 30.59$ m

output power = (7.00×9.8066)N/s $\times$ 30.59 m = 2,100 W
Required motor power = 2.100/0.85 = 2.47 kW

The correct answer is B).

60. The negative pressure at the pump intake equals $5/76 \times 101{,}300 = 6{,}664$ Pa (76 cm Hg equals one atmosphere). The pressure differential across the pump equals $165 + 6.66 = 171.66$ kPa which equals--

$171{,}660/(1{,}000 \times 9.8066) = 17.505$ m

mass flow rate = $(1{,}200/60) \times 1.00 = 20.0$ kg/sec

pump output power = $20.0 \times 9.8066 \times 17.505 = 3{,}433$ W

Efficiency = 3.433/6.3 = 54.5%

The correct answer is A).